Lecture Notes in Biomathematics

Managing Editor: S. Levin

59

Ingemar Nåsell

Hybrid Models of Tropical Infections

Springer-Verlag Berlin Heidelberg GmbH

Author

Ingemar Nåsell
Department of Mathematics, The Royal Institute of Technology
100 44 Stockholm, Sweden

Mathematics Subject Classification (1980): 92A15

ISBN 978-3-540-15978-0 ISBN 978-3-662-01609-1 (eBook)
DOI 10.1007/978-3-662-01609-1

Originally published by Springer-Verlag Berlin Heidelberg New York Tokyo in 1985
Softcover reprint of the hardcover 1st edition 1985

2146/3140-543210

PREFACE

These notes are an extended version of lectures given in the Symposium on Mathematics and Development arranged by the School of Mathematical Sciences of the University of Khartoum, Sudan, in 1982. The purpose of the notes is to discuss some models for the transmission of tropical infections. This area of mathematical epidemiology has previously received only minor attention by mathematicians, but is now growing in importance.

The term "hybrid model" is used to denote a model with both stochastic and deterministic ingredients. We describe how a hybrid model approach can be used to formulate and study both some classical models for malaria and schistosomiasis and some extensions of these models. The formulation of the models requires some familiarity with Markov chains in continuous time and discrete state space. The analysis of the models uses concepts and methods in the qualitative theory of ordinary differential equations. The presentation is aimed at the senior undergraduate or beginning graduate level.

I wish to thank Professor Mohamed E A El Tom of the School of Mathematical Sciences at the University of Khartoum for inviting me to give the lectures on which these notes are based. This work has been supported by the Swedish Natural Sciences Research Council.

ABLE OF CONTENTS

CHAPTER 1. INTRODUCTION

Mathematical epidemiology is a well-established discipline in biomathematics. It is concerned with formulating and studying mathematical models for the transmission of infectious diseases. In a slightly broader framework we can view this work as a study of the ecology of the disease agents. There are two types of infections that differ in their transmission mechanisms in an important way. The first type is exemplified by the well-known viral and bacterial diseases where the infection is transmitted directly from infective to susceptible individuals. The second type is represented by the parasitic diseases that are endemic in many parts of the tropical areas of the world and that are our object of study. Many of the parasitic diseases are caused by parasites that require two or more hosts for the completion of their life cycle, and this is the case for all diseases considered here. A noteworthy consequence is that a human being who is infected by parasites cannot transmit the infection directly to another human being. Instead, the infection is first transmitted to an alternate host who in turn is capable of infecting a human being.

A majority of the work in mathematical epidemiology has been devoted to directly transmitted diseases. This is well testified by Bailey's textbook (1975), where only one out of the 21 chapters is devoted specifically to parasitic diseases. Indeed, work on mathematical models for indirectly transmitted diseases was carried on at a very low level over several decades while during the same period a large number of authors dealt with models for directly transmitted diseases. One reason for this may be that the mathematicians have been unaware of the challenging problems in the area of parasitic infections. The complexity of the transmission patterns certainly requires the mathematician to become more deeply involved in biological and epidemiological phenomena. However, the effort required for this involvement is quite nominal and should not serve as a deterrent. This has also been realized by a number of authors, and a recent trend toward increased work on models for parasitic infections can clearly be discerned. A reflection of this is the appearance of a monograph by Bailey (1982) on mathematical models for malaria.

Malaria and schistosomiasis are the two tropical diseases that have received most attention from model builders. We devote Chapters 3 and 5 to a discussion of several models for the infections corresponding to these two diseases. In addition, a treatment is given of models for so-called hermaphroditic helminthiasis in Chapter 4. We proceed to describe the biological background for these diseases.

1.1 Biological Background

The diseases that we deal with are all caused by parasites with complex life cycles. Different forms of the parasite appear in the different phases of the life cycle. The mathematical models that we establish all reflect this fact. We proceed to give a brief description of the life cycles of the causative parasites for the three infections.

A detailed description of the epidemiology of malaria is given by Bailey (1982). Briefly, the disease is caused by a small protozoan (single-celled) parasite that invades red blood cells of human beings. Inside the blood cells the parasites reproduce asexually. Periodically, the invaded blood cells burst and release parasite forms called merozoites into the blood stream. This causes the recurring fever and chill of infected individuals. The released merozoites proceed to invade new red blood cells; thus the recurring cycle inside the human being begins again.

The parasites belong to the genus *Plasmodium*. There are three species of parasites that infect man. *P. falciparum* is the most virulent one. It is mainly confined to the tropics. *P. vivax* is typical of more temperate zones. Both of these have a period of 48 hours in the human body. The third species, *P. malariae*, has a period of 72 hours.

The asexual cycle serves to multiply the number of parasites in the body of the human host. Some of the merozoites undergo a development that results in a different parasite form called gametocyte. Some of the gametocytes are male and the others are female. Mating between the gametocytes constitutes a sexual phase of reproduction. Such mating does not take place in the human body, but rather in the body of a mosquito. If an infected human being is bitten by a mosquito of the genus *Anopheles* then it can happen that the blood ingested by the mosquito contains both male and female gametocytes. In that case, the gametocytes develop further and mate to produce still another form of the parasite, the oöcyst. It attaches itself to the stomach of the mosquito. The oöcysts in turn give rise to a large number of sporozoites in the salivary glands of the mosquito. When the mosquito bites a human being, the sporozoites are injected into his blood. The sporozoites then develop into merozoites and the life cycle of the parasite is closed.

Schistosomiasis is caused by parasitic flatworms of the genus *Schistosoma*. The disease is also known by the name bilharzia (from Bilharz who in 1851 recovered schistosome worms during autopsy of a patient in Cairo) and snail fever (from the snails that serve as intermediate hosts for the transmission of the infection). There are three species of schistosomes that infect man: *S. mansoni*, *S. haematobium*, and *S. japonicum*. Sexually mature forms of the parasite are found in blood vessels of the intestines of human beings (for *S. mansoni* and *S. japonicum*) or in blood vessels around the urinary bladder (for *S. haematobium*). Male and female forms are paired and the mated female lays fertilized eggs. Some of the eggs are carried by the blood until they become trapped in host tissue. Others succeed in penetrating the wall of the intestine or the bladder and leave the body of the host with the feces or the urine. Some of the eggs that leave the body of the host reach fresh water before they die. When this happens, the eggs hatch, and from each egg emerges a small free-swimming larva, called a miracidium.

If during its short life span a miracidium reaches a fresh-water snail of suitable species, it penetrates the snail and initiates an asexual phase of reproduction in the body of the snail. As a result of this, the infected snail releases swarms of a second form of free-swimming larvae, called cercariae, into the water. The miracidia cannot infect human beings, but the cercariae can. If a cercaria comes sufficiently close to a human being, it will penetrate the skin. Thereafter, it grows to adult size, matures sexually, and migrates to the liver. After mating between male and female schistosomes and migration to a blood vessel, egg-laying starts and the life cycle of the parasite is closed. For a more detailed discussion we refer the reader to Jordan and Webbe (1982).

We use the term hermaphroditic helminthiasis to refer to a class of parasitic diseases exemplified by fasciolopsiasis. This disease is caused by parasitic flatworms (helminths) of the species Fasciolopsis buski. Sexually mature forms of this parasite are hermaphroditic, i.e. each parasite has both male and female reproductive organs. The parasites reside in man s small intestine. Eggs are fertilized and laid and some of them leave the body of the host with the feces. If they reach fresh water, miracidia develop and escape. If the miracidium reaches a snail of suitable species, it infects the snail. An infected snail produces cercariae and releases them into the water. On reaching certain kinds of aquatic vegetation the cercaria changes form into an immobile cyst. Man becomes infected by eating these plants. Further details are given by Chandler and Read (1961).

1.2 The Modelling Approach

The models that we are going to study all refer to a particular community with potential definitive and intermediate hosts for the parasites. A main goal of each model is to describe how the infection level in the community is determined by the biological and environmental parameters that describe the ecology of the parasite.

The diseases we deal with have the property in common that the associated infections are chronic. This is in sharp contrast to many viral or bacterial diseases that are truly epidemic in character. One consequence of this is that it is natural to study steady-state (endemic) infection levels rather than transient phenomena.

We approach the establishment of mathematical models for the transmission of infection in two stages. In the first stage we establish models for the infection of individual hosts or for populations of hosts. Most of these models are stochastic; they are discussed in Chapter 2. In the second stage we make use of the host models as "building blocks" in establishing transmission models. This part of the program is carried out in the explicit treatment of malaria, hermaphroditic helminthiasis and schistosomiasis in Chapters 3-5. The transmission models take the form of a collection of Markov chains, or, in some cases, a collection of Markov chains and deterministic functions. By introducing suitable hybrid hypotheses we are led to deterministic systems of nonlinear ordinary differential equations, whose solutions describe the dynamics of the infection in the community. It turns out that all these systems of differential equations possess a threshold phenomenon. This means that there exists a "threshold" in parameter space that can be used to partition all communities into one of two sets: Communities "above threshold" are those that can support an endemic infection level, while any infection is predicted by the model to die out of its own if the community is "below threshold".

In the study of the threshold phenomenon it is highly important that the parameter space be as simple as possible. We introduce a systematic method, called "quasi-dimensional analysis", for reducing the number of parameters necessary to describe the transmission of the infection.

The mathematical results lead us to introduce some concepts of epidemiological relevance. The term "eradication effort" is closely related to the threshold, while the term "control efficiency" is used to indicate, in an epidemilogically meaningful way, the results of a sensitivity analysis of the transmission model.

CHAPTER 2. HOST MODELS

Models for the transmission of infection in a community will be established in two stages. In the first stage, treated in this chapter, we consider models for the infection of individual hosts and also models for the infection of populations of hosts. These models will be combined in the next three chapters to form transmission models. It is only the latter type of model that possesses threshold phenomena that allow a study of conditions that lead to eradication of the infection.

The host models are built on assumptions about how infections are acquired and lost when the hosts are exposed to a specified environment. The rate at which infections are acquired (the infection rate) may depend both on the infectivity of the environment and on the host's immunity. In other words, the infection rates reflect both the exposure and the susceptibility of the hosts.

Sections 2.1 - 2.5 treat stochastic models for individual hosts, while Section 2.6 deals with stochastic models for a population of hosts. In Section 2.7 we discuss deterministic models for a population of hosts with latency.

2.1 An Infection-Recovery Process

The host in this model has a strong immune reaction when infected but none when uninfected. Only two states of the host are recognized, namely the state in which the host is infected and at the same time immune to additional infections, and the state in which the host is uninfected and nonimmune and therefore susceptible to infection. Two state transitions are posssible, namely infection of an uninfected host and recovery of an infected host. A recovered host is assumed to be immediately susceptible to new infections.

The host model dealt with here is a very simple stochastic process with only two states. The mathematical treatment sets the stage for the approaches to more complicated host models in later sections.

The infection-recovery process will be used as host model for both the human and mosquito phases of infection in the Ross malaria model of Section 3.1, for the mosquito phase of the malaria model with superinfection of Section 3.2, for the snail phases of the two models for hermaphroditic helminthiasis in Chapter 4, and for the snail phases for two of the models for schistosomiasis of Chapter 5, namely those treated in Sections 5.1 and 5.4.

We introduce the Markov chain $\underline{X}(\underline{t})$, $\underline{t} \geq 0$, to indicate the state of the host at time $\underline{t}$. The state space is $\{0,1\}$, where the state 0 corresponds to an uninfected host, and the state 1 corresponds to an infected host. The transition probabilities are denoted by

$$P_{mn}(s,t) = P\{X(t) = n | X(s) = m\}, \quad m,n = 0,1, \quad 0 \leq s \leq t. \qquad (2.1.1)$$

The infection rate is denoted by $\underline{h}(\underline{t})$, where $\underline{h}$ is a nonnegative continuous function. The recovery rate is denoted by $\underline{r}$, a positive constant. The hypotheses of the model are formulated in terms of expressions for the transition probabilities over short time intervals. The hypotheses follow from a strict interpretation of the rates of infection and recovery, as follows:

$$P_{01}(t,t+\Delta t) = h(t)\Delta t + o(\Delta t), \qquad (2.1.2)$$

$$P_{00}(t,t+\Delta t) = 1 - h(t)\Delta t + o(\Delta t), \qquad (2.1.3)$$

$$P_{10}(t,t+\Delta t) = r\Delta t + o(\Delta t), \qquad (2.1.4)$$

$$P_{11}(t,t+\Delta t) = 1 - r\Delta t + o(\Delta t), \quad \Delta t \to 0. \qquad (2.1.5)$$

The expression $\underline{o}(\Delta\underline{t})$ denotes a function which, when divided by $\Delta\underline{t}$, approaches zero as $\Delta\underline{t}$ tends to zero.

Equation (2.1.2) shows that the probability of infection of an uninfected host in a short time interval is essentially equal to the product of the infection rate and the length of the time interval. Similarly, equation (2.1.4) shows that the probability of recovery of an infected host in a short time interval is essentially equal to the product of the recovery rate and the length of the time interval.

The Markov property specifies that if the present state of the process is given, then the future behaviour is independent of the past. Consider three consecutive epochs of time $\underline{s} < \underline{t} < \underline{t}+\Delta\underline{t}$. If the process is in state 0 at time $\underline{s}$ and in state 1 at time $\underline{t}+\Delta\underline{t}$, then it must have been in either state 0 or state 1 at time $\underline{t}$. The probability of the first of these two paths (the one going through state 0 at time $\underline{t}$) is

$$P\{X(t+\Delta t)=1,\ X(t)=0|X(s)=0\} =$$

$$= P\{X(t)=0|X(s)=0\}\cdot P\{X(t+\Delta t)=1|X(t)=0,\ X(s)=0\}. \quad (2.1.6)$$

The first of the two conditional probabilities on the right hand side of this expression is by definition equal to the transition probability $P_{00}(\underline{s},\underline{t})$, while the second one by the Markov property is equal to $P_{01}(\underline{t},\underline{t}+\Delta\underline{t})$. By similar arguments we find the probability of the second path to be $P_{01}(\underline{s},\underline{t})P_{11}(\underline{t},\underline{t}+\Delta\underline{t})$. The probability for the process to be in state 1 at time $\underline{t}+\Delta\underline{t}$, given that it is in state 0 at time $\underline{s}$, is the sum of these two path probabilities, or

$$P_{01}(s,t+\Delta t) = P_{00}(s,t)P_{01}(t,t+\Delta t) + P_{01}(s,t)P_{11}(t,t+\Delta t). \quad (2.1.7)$$

We insert expressions for the infinitesimal transition probabilities on the right hand side of this expression from (2.1.2) and (2.1.5), use the relation

$$P_{00}(s,t) + P_{01}(s,t) = 1, \quad (2.1.8)$$

make a simple rearrangement, divide by Δt, and let Δt approach zero. We then find that $P_{01}(s,t)$ statisfies the differential equation

$$\frac{\partial}{\partial t} P_{01}(s,t) = h(t) - (h(t)+r)P_{01}(s,t). \qquad (2.1.9)$$

In a similar manner we find that the transition probability function $P_{10}(s,t)$ satisfies the differential equation

$$\frac{\partial}{\partial t} P_{10}(s,t) = r - (h(t)+r)P_{10}(s,t). \qquad (2.1.10)$$

These are the so-called forward Kolmogorov differential equations, since differentiation is with respect to the later time t. They are partial differential equations since P_{01} and P_{10} depend on both s and t. However, they can be treated as ordinary differential equations in t for each fixed value of s. Initial conditions are given for $t = s$ as follows:

$$P_{01}(s,s) = 0, \qquad (2.1.11)$$

$$P_{10}(s,s) = 0. \qquad (2.1.12)$$

The initial value problems posed for the functions P_{01} and P_{10} by the above equations and initial conditions can be solved explicitly. The differential equations are linear with time-dependent coefficients. We shall encounter such equations repeatedly in this chapter. For convenience in the treatment of such equations we consider a differential equation of the following general form:

$$x' = a(t) - b(t)x, \qquad t \geq s. \qquad (2.1.13)$$

We assume that the initial value of x, $x(s)$, is specified at $t = s$. We introduce two functions A and B by putting

$$B(t) = \int_0^t b(u)du \qquad (2.1.14)$$

and

$$A(t) = e^{-B(t)} \int_0^t a(v)e^{B(v)} dv \, . \tag{2.1.15}$$

After some elementary manipulations we find that the solution of equation (2.1.13) can be written in the form

$$x(t) = A(t) + (x(s) - A(s))e^{-(B(t)-B(s))} \, . \tag{2.1.16}$$

This general result can be used to find the solutions $\underline{P}_{01}$ and $\underline{P}_{10}$ of the initial value problems posed by (2.1.9) - (2.1.12). We define two functions τ and σ as follows:

$$\tau(t) = rt + \int_0^t h(u)du, \tag{2.1.17}$$

$$\sigma(t) = r\, e^{-\tau(t)} \int_0^t e^{\tau(u)} du. \tag{2.1.18}$$

The solutions can then be written

$$P_{01}(s,t) = 1-\sigma(t) - (1-\sigma(s))e^{-(\tau(t)-\tau(s))}, \tag{2.1.19}$$

$$P_{10}(s,t) = \sigma(t) - \sigma(s)e^{-(\tau(t)-\tau(s))} \, . \tag{2.1.20}$$

The probability $\underline{p}(\underline{t})$ that the host is infected at time $\underline{t}$ is

$$p(t) = P\{X(t) = 1\} \, . \tag{2.1.21}$$

It is found from the above expressions for the transition probabilities and the initial infection probability $\underline{p}(0)$ that

$$p(t) = p(0)P_{11}(0,t) + (1-p(0))P_{01}(0,t) =$$

$$= p(0)e^{-\tau(t)} + 1-\sigma(t). \tag{2.1.22}$$

This is an explicit expression for the infection probability $\underline{p}(\underline{t})$ in terms of the initial infection probability $\underline{p}(0)$ and the functions τ and σ , whose values depend on the given infection rate $h(t)$ and the given recovery rate r. For the formulation of transmission models we shall find it convenient to make use of a differential equation satisfied by the infection probability $p(t)$. Differentiation of (2.1.22) gives

$$p'(t) = h(t)(1 - p(t)) - rp(t). \qquad (2.1.23)$$

We proceed to evaluate the three epidemiological quantities prevalence, incidence, and recovery probability. The <u>prevalence</u> $Q(s)$ is defined as the probability that an individual host of age s is infected. The <u>incidence</u> $I(s,T)$ is the probability that an uninfected host of age s will be infected a fixed time T later. The <u>recovery probability</u> $R(s,T)$ is the probability that an infected host of age s will be free from infection a fixed time T later. The evaluations of these quantities will be made for steady-state epidemiological situations where the infection rate $h(t)$ equals a nonnegative constant. We denote this constant by H. Under this restriction, the epidemiological quantities do not depend on the time t .

The age-dependence of prevalence, incidence and recovery probability are found by considering a cohort of individuals exposed to the environment and studying their acquisition and loss of infection as a function of time. The initial age of the individuals in the cohort is taken to be zero, and newborn individuals are assumed to be free from infection. The elapsed time is then equal to the age s of the individuals in the cohort. From the definitions of prevalence, incidence and recovery probability we find

$$Q(s) = P_{01}(0,s), \qquad (2.1.24)$$

$$I(s,T) = P_{01}(s,s+T), \qquad (2.1.25)$$

$$R(s,T) = P_{10}(s,s+T), \qquad (2.1.26)$$

where the transition probabilities are to be evaluated for the special case where the infection rate is constant, i.e. $h(t) = H$.

The functions τ and σ defined in (2.1.17) and (2.1.18) are then found to be as follows:

$$\tau(t) = (H + r)t, \qquad (2.1.27)$$

$$\sigma(t) = \frac{r}{H+r}\left(1 - e^{-(H+r)t}\right). \qquad (2.1.28)$$

By using the expressions (2.1.19) and (2.1.20) for the transition probability functions we get

$$Q(s) = \frac{H}{H+r}\left(1 - e^{-(H+r)s}\right), \qquad (2.1.29)$$

$$I(s,T) = Q(T), \qquad (2.1.30)$$

$$R(s,T) = \frac{r}{H} Q(T). \qquad (2.1.31)$$

For the simple model treated in this section we thus find that the prevalence increases monotonically with age toward the value $\underline{H}/(\underline{H}+r)$. Furthermore, the incidence and the recovery probability are both independent of age. They increase monotonically with the length $\underline{T}$ of the time interval over which they are defined toward the values $\underline{H}/(\underline{H}+\underline{r})$ and $\underline{r}/(\underline{H}+\underline{r})$, respectively.

2.2 A Superinfection Process

This section treats a stochastic model which will be used as host model for the human phase of the infection in all three diseases treated in Chapter 3, namely malaria, hermophroditic helminthiasis and schistosomiasis. In each case, the model accounts for the number of infections in the host. Each infection corresponds to a brood of parasites in malaria with superinfection and to a living parasite in hermaphroditic helminthiasis and schistosomiasis. The superinfection process is equivalent to a nonhomogeneous immigration-death process for the number of infections in the host.

We introduce the Markov chain $\underline{X}(\underline{t})$, $\underline{t} \geq 0$, to indicate the number of infections in the host at time $\underline{t}$. The state space is the set of nonnegative integers $\{0,1,2,...\}$. The transition probabilities are denoted $\underline{P}_{mn}(\underline{s},\underline{t})$, $0 \leq \underline{s} \leq \underline{t}$, where now $\underline{m}$ and $\underline{n}$ take values in the state space, i.e. $\underline{m},\underline{n} = 0,1,2,...$. The hypotheses of the process are based on a nonnegative continuous function $\underline{h}(\underline{t})$, $\underline{t} \geq 0$, and on a positive constant $\underline{r}$. By using the language of immigration-death processes, we refer to $\underline{h}(\underline{t})$ as the immigration rate and to $\underline{r}$ as the death rate per individual. For our applications, the immigration rate $\underline{h}(\underline{t})$ can be interpreted as the infection rate per host, while the parameter $\underline{r}$ is interpreted as the death rate per parasite for hermaphroditic helminthiasis and schistosomiasis and as the rate of recovery per individual infection for malaria with superinfection. We proceed to formulate the hypotheses of the model by specifying the infinitesimal transition probabilities as $\Delta\underline{t} \to 0$:

$$P_{n,n+1}(t,t+\Delta t) = h(t)\Delta t + o(\Delta t), \quad n=0,1,... \tag{2.2.1}$$

$$P_{n,n-1}(t,t+\Delta t) = nr\Delta t + o(\Delta t), \quad n=1,2,... \tag{2.2.2}$$

$$P_{n,n}(t,t+\Delta t) = 1 - (h(t) + nr)\Delta t + o(\Delta t), \; n=0,1,... \tag{2.2.3}$$

Equation (2.2.1) shows that the probability for a host of acquiring an additional infection in a short time interval is essentially equal to the product of the immigration rate and the length of the time interval. The probability for a given infection to disappear in a short time interval is essentially equal to the product of the death rate and the length of the time interval. The event that the number of infections in a host decreases by one during a short time interval is equal to the event that one of the infections disappears during the time interval. The probability of this event, given in (2.2.2), is essentially equal to the product of the number of infections at the beginning of the time interval, the death rate, and the length of the time interval.

Our first goal is to find explicit expressions for the transition probability functions that satisfy (2.2.1)-(2.2.3) and the obvious initial condition

$$P_{mn}(s,s) = \begin{cases} 1, & n=m, \\ 0, & n\neq m. \end{cases} \tag{2.2.4}$$

One standard approach to this problem is to use the hypotheses and the Chapman-Kolmogorov equations to derive a system of differential equations (the Kolmogorov forward equations) for the transition probability functions $P_{mn}(s,t)$. One then introduces the probability generating function

$$G(t,z;s,m) = \sum_{n=0}^{\infty} P_{mn}(s,t)z^n \tag{2.2.5}$$

and uses the system of differential equations to derive a partial differential equation for G, holding s and m fixed. Finally, one solves the partial differential equation for G and recovers the transition probability functions.

Bailey (1964, Section 7.4) has described a so-called random-variable technique that considerably simplifies the derivation of the partial differential equation for the probability generating function G. We shall follow his approach. Bailey's technique is based on the assumption that only a finite number of transitions during a time interval of length Δt have transition probabilities of the order of Δt. For these transitions Bailey introduces the notation

$$P_{n,n+j}(t,t+\Delta t) = f_j(n)\Delta t + o(\Delta t), \quad \Delta t \to 0, \quad j \neq 0. \tag{2.2.6}$$

We note from (2.2.1)-(2.2.3) that only transitions with $j = +1$ and $j = -1$ appear for the superinfection process. Bailey then shows that the probability generation function satisfies the partial differential equation

$$\frac{\partial G(t,z)}{\partial t} = \sum_{j\neq 0} (z^j-1)f_j(z\frac{\partial}{\partial z})G(t,z). \tag{2.2.7}$$

From (2.2.1) and (2.2.2) we find that

$$f_1(n) = h(t) \tag{2.2.8}$$

and that

$$f_{-1}(n) = nr. \tag{2.2.9}$$

Substitution into (2.2.7) leads to the partial differential equation

$$\frac{\partial G}{\partial t} + r(z-1)\frac{\partial G}{\partial z} = h(t)(z-1)G. \tag{2.2.10}$$

The initial data for G are specified in the t-z-plane along the line $t = s$ by

$$G(s,z;s,m) = z^m, \tag{2.2.11}$$

which follows from the initial conditions (2.2.4) for the transition probability functions.

The partial differential equation (2.2.10) is a linear first-order equation that can be solved by using the theory of characteristics. This approach involves the derivation of an ordinary differential equation for a closely related function $\bar{G}$, which is a function of t alone. Along any curve $\bar{z}(t)$ in the t-z -plane we define

$$\bar{G}(t;s,m) = G(t,\bar{z}(t);s,m). \tag{2.2.12}$$

The derivative of $\bar{G}$ with respect to t is clearly related to the partial derivatives of G with respect to t and z as follows:

$$\bar{G}' = \frac{\partial G}{\partial t} + \frac{\partial G}{\partial z}\bar{z}'(t). \tag{2.2.13}$$

The curve $\bar{z}(t)$ is a characteristic for the partial differential equation (2.2.10) if $\bar{z}$ satisfies the ordinary differential equation

$$\bar{z}'(t) = r(\bar{z}(t) - 1). \tag{2.2.14}$$

If $\bar{z}(t)$ is a characteristic, then a comparison between (2.2.10) and (2.2.13) shows that the function $\bar{G}$ satisfies the ordinary differential equation

$$\bar{G}' = h(t)(\bar{z}(t) - 1)\bar{G}. \qquad (2.2.15)$$

Now any characteristic, i.e. any solution of the equation (2.2.14), can be written in the form

$$\bar{z}(t) = 1 + Ke^{rt}, \qquad (2.2.16)$$

where K is a constant. Thus the differential equation for $\bar{G}$ along the corresponding characteristic becomes

$$\bar{G}' = h(t)Ke^{rt}\bar{G} . \qquad (2.2.17)$$

The initial value for $\bar{G}$ is specified at $t = s$ by

$$\bar{G}(s;s,m) = G(s,\bar{z}(s);s,m) = G(s,1+Ke^{rs};s,m) =$$

$$= (1 + Ke^{rs})^m, \qquad (2.2.18)$$

as we find from the definition of $\bar{G}$ in (2.2.12), the expression for the characteristic $\bar{z}(t)$ in (2.2.16), and the initial data for G given in (2.2.11).

The initial value problem posed for the function $\bar{G}$ by (2.2.17) and (2.2.18) can be solved explicitly. The solution is

$$\bar{G}(t;s,m) = (1 + Ke^{rs})^m \exp(K(e^{rt}\gamma(t) - e^{rs}\gamma(s))), \qquad (2.2.19)$$

where the function γ is defined by

$$\gamma(t) = e^{-rt} \int_0^t h(u)e^{ru}du. \qquad (2.2.20)$$

Expression (2.2.19) gives $\bar{G}$ as a function of t on the characteristic determined by K.

The solution to the initial value problem posed by (2.2.10), (2.2.11) for the function G can now be found as follows. The function value of $\underline{G}$ for given values of $\underline{t}$ and $\underline{z}$ is the value that the function $\bar{\underline{G}}$ takes for the same value of $\underline{t}$ on the characteristic passing through the point $(\underline{t},\underline{z})$. The $\underline{K}$-value determining this characteristic is found from (2.2.16) to equal

$$K = (z-1)e^{-rt}. \qquad (2.2.21)$$

Insertion of this value into (2.2.19) gives the following expression for $\underline{G}$:

$$G(t,z;s,m) = (1+(z-1)q(s,t))^{m} \exp((z-1)\lambda(s,t)), \qquad (2.2.22)$$

where

$$q(s,t) = e^{-r(t-s)} \qquad (2.2.23)$$

and

$$\lambda(s,t) = \gamma(t) - q(s,t)\gamma(s). \qquad (2.2.24)$$

It remains to recover explicit expressions for the transition probability functions $\underline{P}_{mn}(\underline{s},\underline{t})$ from this result. To do this we note first the elementary results, which are easily verified from (2.2.5), that the first factor on the right-hand side of (2.2.22) is the probability generating function of a binomially distributed random variable $\underline{X}_1(\underline{t})$ with parameters $\underline{m}$ and $\underline{q}(\underline{s},\underline{t})$, and that the second factor on the right-hand side of (2.2.22) is the probability generating function of a Poisson distributed random variable $\underline{X}_2(\underline{t})$ with parameter $\lambda(\underline{s},\underline{t})$. Next we note that the probability generating function of a sum of two independent random variables equals the product of the latter's probability generating functions. For fixed values of $\underline{m}$, $\underline{s}$ and $\underline{t}$ the transition probability function $\underline{P}_{mn}(\underline{s},\underline{t})$ gives the probability distribution of $\underline{X}(\underline{t})$, conditional on the event that $\underline{X}(\underline{s}) = \underline{m}$.

Thus, conditioning on this event, we conclude that $\underline{X}(\underline{t})$ can be written as the sum of the two independent random variables $\underline{X}_1(\underline{t})$ and $\underline{X}_2(\underline{t})$. These random variables can be interpreted as the number of surviving natives and surviving immigrants, respectively. The probability distribution $\{\underline{P}_{mn}(\underline{s},\underline{t})\}$, where $\underline{n} = 0,1,\ldots,$ can be written as the convolution of the distributions of $\underline{X}_1(\underline{t})$ and $\underline{X}_2(\underline{t})$. The resulting expressions for the transition probability functions are:

$$P_{mn}(s,t) =$$

$$= \exp(-\lambda(s,t)) \sum_{j=0}^{m\wedge n} \binom{m}{j} (q(s,t))^j (1-q(s,t))^{m-j} \frac{(\lambda(s,t))^{n-j}}{(n-j)!} , \tag{2.2.25}$$

where we use the notation

$$m\wedge n = \min(m,n). \tag{2.2.26}$$

Note that the summation over j can be extended to m if we adopt the convention that $1/n! = 0$ for $n < 0$. For use in the sequel we note in particular that

$$P_{00}(s,t) = \exp(-\lambda(s,t)). \tag{2.2.27}$$

The probability distribution of the number of infections at time t is denoted $p_n(t)$, i.e.

$$p_n(t) = P\{X(t) = n\}, \quad n = 0,1,2,\ldots , \quad t \geq 0. \tag{2.2.28}$$

It is found from the transition probability functions and the initial distribution p_m through the relation

$$p_n(t) = \sum_{m=0}^{\infty} p_m P_{mn}(0,t). \tag{2.2.29}$$

If the Markov chain in the model is homogeneous, i.e. if the immigration rate $h(t)$ is constant, then one can show that $X(t)$ has a stationary Poisson distribution. Thus, if the stationary distribution is taken as initial distribution, then the distribution of $X(t)$ is the same for all $t \geq 0$. The same result can clearly not hold for the nonhomogeneous Markov chain treated in this section. However, we shall establish a certain generalization of this result. In particular, we shall prove that if the initial distribution is Poisson with an arbitrary parameter x_0, then the distribution of $X(t)$ is Poisson for each $t \geq 0$ with a time-varying parameter.

In order to prove this result we condition first on the event $X(0) = m$. As noted above, $X(t)$ is then the sum of two independent random variables $\underline{X}_1(\underline{t})$ and $\underline{X}_2(\underline{t})$, where now $\underline{X}_1(\underline{t})$ has a binomial distribution with parameters $\underline{m}$ and $\underline{e}^{-rt}$, and $\underline{X}_2(\underline{t})$ has a Poisson distribution with parameter $\gamma(\underline{t})$. The distribution of $\underline{X}_2(\underline{t})$ does not depend on $\underline{m}$, and is therefore unaffected by a removal of the conditioning. For the distribution of $\underline{X}_1(\underline{t})$ we find by taking the expectation with respect to the initial distribution that

$$P\{X_1(t)=k\} = \sum_{m=0}^{\infty} \frac{x_0^m}{m!} e^{-x_0} \binom{m}{k} e^{-krt} (1-e^{-rt})^{m-k}$$

$$= \frac{1}{k!} (x_0e^{-rt})^k e^{-x_0} \sum_{i=0}^{\infty} \frac{(x_0(1-e^{-rt}))^i}{i!} = \frac{1}{k!} (x_0e^{-rt})^k e^{-x_0e^{-rt}}. \tag{2.2.30}$$

This expression shows that $\underline{X}_1(\underline{t})$ has a Poisson distribution with parameter $\underline{x}_0e^{-rt}$. Thus, $\underline{X}(\underline{t})$ is the sum of two independent Poisson-distributed random variables. It follows that $\underline{X}(\underline{t})$ itself has a Poisson distribution, and that its parameter $x(t)$ is the sum of the parameters of $\underline{X}_1(\underline{t})$ and $\underline{X}_2(\underline{t})$, as follows:

$$x(t) = EX(t) = x_0e^{-rt} + \gamma(t). \tag{2.2.31}$$

The formulation of transmission models with the host model of the present section as a building block requires a differential equation for the expected number $\underline{x}(\underline{t})$ of infections at time $\underline{t}$. Differentiation of (2.2.31) gives

$$x' = -rx_0e^{-rt} + \gamma'(t) = h(t) - rx. \tag{2.2.32}$$

We proceed to evaluate the prevalence $\underline{Q}(\underline{s})$, the incidence $\underline{I}(\underline{s},\underline{T})$ and the recovery probability $\underline{R}(\underline{s},\underline{T})$ in steady state, i.e. in the case where the immigration rate is constant, $\underline{h}(\underline{t}) = \underline{H}$. The approach uses a cohort as described in the previous section. The prevalence and the incidence are readily expressed in terms of the distribution and the transition probability functions as follows:

$$Q(s) = P\{X(s) > 0\} = 1 - p_0(s), \tag{2.2.33}$$

$$I(s,T) = P\{X(s+T) > 0 \mid X(s) = 0\} = \sum_{n=1}^{\infty} P_{0n}(s,s+T)$$

$$= 1 - P_{00}(s,s+T). \qquad (2.2.34)$$

For the recovery probability we have by definition

$$R(s,T) = P\{X(s+T) = 0 | X(s) > 0\} . \qquad (2.2.35)$$

The simplest way to derive an expression for the recovery probability is to use the following relation between prevalence, incidence, and recovery probability:

$$Q(s+T) = Q(s)(1 - R(s,T)) + (1 - Q(s))I(s,T). \qquad (2.2.36)$$

By solving for the recovery probability we get

$$R(s,T) = 1 - I(s,T) - \frac{Q(s+T) - I(s,T)}{Q(s)} . \qquad (2.2.37)$$

For the cohort treated all individuals of age zero are free from infection. Therefore the following relation holds between the probability of being free from infection at age $\underline{s}$ and the transition probability function value $\underline{P}_{00}(0,\underline{s})$:

$$p_0(s) = P_{00}(0,s). \qquad (2.2.38)$$

In the present case with constant immigration rate $\underline{h}(\underline{t}) = \underline{H}$ we find from (2.2.20) that

$$\gamma(t) = \frac{H}{r}(1 - e^{-rt}). \qquad (2.2.39)$$

It follows therefore from (2.2.27) that the transition probability function $\underline{P}_{00}$ is given by

$$P_{00}(s,t) = \exp(-\frac{H}{r}(1 - e^{-r(t-s)})). \qquad (2.2.40)$$

We note that the function value depends only on the difference, $\underline{t}$-$\underline{s}$, between the arguments. A consequence of this and relation (2.2.38) is that the incidence in (2.2.34) can be expressed as follows:

$$I(s,T) = 1 - p_0(T). \tag{2.2.41}$$

Furthermore, it follows from (2.2.33), (2.2.41) and (2.2.37) that the recovery probability equals

$$R(s,T) = p_0(T) - \frac{p_0(T)-p_0(s+T)}{1 - p_0(s)} = \frac{p_0(s+T)-p_0(s)p_0(T)}{1 - p_0(s)} . \tag{2.2.42}$$

Thus, prevalence, incidence, and recovery probability can all be expressed in term of values of the function p_0. By using (2.2.38) and (2.2.40) we find that this function is given by the expression

$$p_0(t) = \exp\left(- \frac{H}{r}\,\alpha(t)\right) , \tag{2.2.43}$$

where the function α is defined by

$$\alpha(t) = 1 - e^{-rt}. \tag{2.2.44}$$

From this expression we conclude that $p_0(t)$ decreases monotonically in t toward the value $e^{-H/r}$ as $t \to \infty$. It follows from (2.2.33) that the prevalence $Q(s)$ increases monotonically with the age s toward the value $1 - e^{-H/r}$. Expression (2.2.41) shows that the incidence is independent of the age s, and that it increases monotonically with the length T of the time interval over which it is defined toward the value $1 - e^{-H/r}$. The recovery probability $R(s,T)$ is a monotonically decreasing function of the age s. The reason for this is that infected old individuals tend to have a larger number of infections than infected young ones, and that recovery requires termination of all the infections. The dependence of the recovery probability on the length T of the time interval over which it is defined is more complicated. It turns out that $R(s,T)$ increases monotonically with T if the infection level is small or if the age is high, but that it has a maximum as function of T if the infection level is high and the age is small. Details are given in Appendix I.

The probability for an individual host to recover during a short time interval can be used to define a host recovery rate $R_1(t)$ as follows:

$$R(t,\Delta t) = R_1(t)\Delta t + o(\Delta t), \quad \Delta t \to 0. \tag{2.2.45}$$

The left-hand side of this expression can be written

$$P\{X(t+\Delta t) = 0 | X(t) > 0\} = \frac{P\{X(t)=1\} P_{10}(t, t+\Delta t)}{P\{X(t) > 0\}} + o(\Delta t), \tag{2.2.46}$$

since $\underline{P}_{n0}(\underline{t}, \underline{t} + \Delta \underline{t}) = o(\Delta \underline{t})$, $\underline{n} > 1$. By using the expression (2.2.2) for the transition probability $\underline{P}_{10}(\underline{t}, \underline{t} + \Delta \underline{t})$ we get the following expression for the host recovery rate $\underline{R}_1(\underline{t})$:

$$R_1(t) = \frac{rP\{X(t) = 1\}}{P\{X(t) > 0\}}. \tag{2.2.47}$$

Under the assumption that the initial distribution is Poisson we know that $\underline{X}(\underline{t})$ has a Poisson distribution whose mean value $\underline{x}(\underline{t})$ is given by (2.2.31). It follows then that

$$R_1(t) = \frac{rx(t)e^{-x(t)}}{1 - e^{-x(t)}}. \tag{2.2.48}$$

The recovery rate $\underline{R}_1(\underline{t})$ can also be expressed in terms of the rate $\underline{r}$ of termination of an individual infection and the infection probability $\underline{p}(\underline{t}) = \underline{P}\{\underline{X}(\underline{t}) > 0\}$. We get

$$R_1(t) = r \frac{1 - p(t)}{p(t)} \ln \frac{1}{1 - p(t)}. \tag{2.2.49}$$

This relation is analyzed further in the discussion of the influence of superinfection in malaria in Subsection 3.2.7.

2.3 A Superinfection Process with Immunity

The infection-recovery process treated in Section 2.1 is built on an assumption of a strong immune reaction of the infected host. We proceed here to formulate a host model where the immune reaction of the infected host is weaker. Human immunity to many parasitic infections has the properties that it develops gradually, that it is only partial, and that it does not (as in the model of Section 2.1) prevent superinfection to take place.

Mathematical models can be formulated in a number of ways to reflect these properties. First of all, it is necessary to define a quantity that measures the level of immunity. In concomitant immunity the immune level is simply equal to the number of infections or the number of living parasites, while in subsequent immunity it is a more complex measure of the host's previous exposure to infections; see the discussion in Cohen (1977). We base the model of the present section on the assumption of concomitant immunity. Secondly, we need to introduce hypotheses that reflect the mechanism by which the immune level affects the infectious process. Two possibilities suggest themselves. An increased immune level can be hypothesized either to lead to a decrease in the infection rate or to an increase in the rate of recovery per infection. We choose the former alternative.

The model to be developed here is an immigration-death process with constant death rate $\underline{r}$ per individual. This parameter can alternatively be described as the rate of recovery per individual infection. The immigration rate (infection rate) is taken to be a linearly decreasing function of the number of infections. To achieve this, we postulate the existence of a positive integral-valued saturation level $\underline{S}$ with the property that the immigration rate is proportional to the difference between the saturation level and the immune level. The immigration rate for a host with no parasites (i.e. with immune level zero) is determined by the infectivity of the environment and is denoted by $\underline{h}(\underline{t})$, as in the previous sections. The immigration rate at time $\underline{t}$ is therefore equal to $(1 - \underline{n}/\underline{S})\underline{h}(\underline{t})$, where $\underline{n}$ is the number of infections in the host at time $\underline{t}$. We use $\underline{X}(\underline{t})$ to denote a Markov chain that corresponds to the number of infections in the host at time $\underline{t}$. The state space is the finite set of integers $0,1,\ldots,\underline{S}$. The transition probabilities of the process are denoted $\underline{P}_{mn}(\underline{s},\underline{t})$. The hypotheses of the model are formulated as follows for $\Delta t \to 0$:

$$P_{n,n+1}(t,t+\Delta t) = (1-n/S)h(t)\Delta t+o(\Delta t),\ n=0,1,\ldots,S-1, \tag{2.3.1}$$

$$P_{n,n-1}(t,t+\Delta t) = nr\Delta t + o(\Delta t),\ \ n=1,2,\ldots,S, \tag{2.3.2}$$

$$P_{n,n}(t,t+\Delta t) = 1 - ((1-n/S)h(t) + nr)\Delta t + o(\Delta t),\ n=0,1,\ldots,S.$$

Initial conditions are given at $t=s$ by

$$P_{mn}(s,s) = \begin{cases} 1, & n = m, \\ 0, & n \neq m. \end{cases} \tag{2.3.4}$$

The probability generating function G is defined as follows:

$$G(t,z;s,m) = \sum_{n=0}^{S} P_{mn}(s,t)z^n. \tag{2.3.5}$$

We proceed to derive a partial differential equation for G as function of t and z, holding s and m fixed. To do this, we apply Bailey's random-variable technique described in the previous section. We note from (2.3.1) and (2.3.2) that

$$f_1(n) = (1 - \tfrac{n}{S})\ h(t) \tag{2.3.6}$$

and that

$$f_{-1}(n) = rn. \tag{2.3.7}$$

It follows from (2.2.7) that G satisfies the partial differential equation

$$\frac{\partial G}{\partial t} + (r + h(t)z/S)(z-1)\ \frac{\partial G}{\partial z} = h(t)(z-1)G. \tag{2.3.8}$$

From the initial conditions (2.3.4) for the transition probability functions we find that initial data for G are given in the t - z-plane along the line $t = s$ by

$$G(s,z;s,m) = z^m. \tag{2.3.9}$$

We proceed to solve the initial-value problem (2.3.8), (2.3.9) for the function G by using the theory of characteristics. A characteristic is a curve $\bar{z}(t)$ that satisfies the ordinary differential equation

$$\bar{z}'(t) = (r + h(t)\bar{z}(t)/S)(\bar{z}(t) - 1). \tag{2.3.10}$$

Along any characteristic $\bar{z}(t)$ we define $\bar{G}$ as a function of t by setting

$$\bar{G}(t;s,m) = G(t,\bar{z}(t);s,m). \tag{2.3.11}$$

It follows from (2.3.8) and (2.3.10) that the function $\bar{G}$ satisfies the ordinary differential equation

$$\bar{G}' = h(t)(\bar{z}(t) - 1)\bar{G} \tag{2.3.12}$$

with the initial condition given at $t = s$ by

$$\bar{G}(s;s,m) = G(s,\bar{z}(s);s,m). \tag{2.3.13}$$

The differential equation (2.3.10) for any characteristic $\bar{z}(t)$ is a Riccati equation which clearly is satisfied by the particular solution $\bar{z}(t) = 1$. A standard way of solving such an equation is to subtract the particular solution from $\bar{z}(t)$ and introduce a new function $y(t)$ as the inverse of the difference. Thus we define

$$y(t) = 1/(\bar{z}(t) - 1). \tag{2.3.14}$$

Differentiation of this expression shows that the function y satisfies a linear differential equation:

$$y' = -h(t)/S - (r + h(t)/S)y. \tag{2.3.15}$$

This equation can be solved explicitly. By using (2.1.16) we find that the solution has the form

$$y(t) = (K - H(t))e^{-\rho(t)}, \tag{2.3.16}$$

where K is a constant and where the functions ρ and H are defined by the relations

$$\rho(t) = rt + \frac{1}{S}\int_0^t h(u)du \tag{2.3.17}$$

and

$$H(t) = \frac{1}{S}\int_0^t h(u)\, e^{\rho(u)} du. \qquad (2.3.18)$$

It follows that any characteristic $\bar{z}(t)$ can be written in the form

$$\bar{z}(t) = 1 - e^{\rho(t)}/(H(t)-K). \qquad (2.3.19)$$

The initial value problem (2.3.12), (2.3.13) for the function $\bar{G}$ can therefore be formulated explicitly as follows:

$$\bar{G}' = -\,(h(t)\, e^{\rho(t)}/(H(t)-K))\bar{G}, \qquad (2.3.20)$$

$$\bar{G}(s;s,m) = (1 - e^{\rho(s)}/(H(s)-K))^m. \qquad (2.3.21)$$

The explicit solution for the function $\bar{G}$ along the characteristic $\bar{z}(t)$ determined by the constant K is found to be

$$\bar{G}(t;s,m) = \left(1 - \frac{e^{\rho(s)}}{H(s)-K}\right)^m \exp\left(-\int_s^t \frac{h(u)e^{\rho(u)}}{H(u)-K}\, du\right). \qquad (2.3.22)$$

The definition (2.3.18) for the function H shows that

$$H'(t) = h(t)e^{\rho(t)}/S. \qquad (2.3.23)$$

By using this relation we find that the integral on the right-hand side of (2.3.22) can be evaluated. We get

$$\int_s^t \frac{h(u)e^{\rho(u)}}{H(u)-K} du = S\int_s^t \frac{H'(u)}{H(u)-K} du = \ln\left|\frac{H(t)-K}{H(s)-K}\right|^S. \qquad (2.3.24)$$

By inserting this expression into (2.3.22) we find that the function $\bar{G}$ can be written in the form

$$\bar{G}(t;s,m) = \left[1 - \frac{e^{\rho(s)}}{H(s)-K}\right]^m \left|\frac{H(s)-K}{H(t)-K}\right|^S \qquad (2.3.25)$$

Thus we have found an expression for the function $\bar{G}$ as a function of t. The corresponding value of the function G for given values of t and z is equal to the value that the function $\bar{G}$ takes for the same value of t on the characteristic passing through the point (t,z). The K-value that determines this characteristic is found from (2.3.19) to be equal to

$$K = H(t) - e^{\rho(t)}/(1-z). \tag{2.3.26}$$

We note with this expression for K that

$$H(t) - K = \frac{e^{\rho(t)}}{1-z} > 0, \quad z < 1 \tag{2.3.27}$$

and that

$$H(s) - K \geq e^{\rho(t)} - H(t), \; 0 \leq z < 1. \tag{2.3.28}$$

The right-hand side of this expression is equal to zero for $t = 0$ and its derivative with respect to t is equal to $re^{\rho(t)}$. We conclude that the right-hand side of (2.3.28) is nonnegative. Insertion of (2.3.26) into (2.3.25) therefore gives the following expression for G, valid for $0 \leq z < 1$:

$$G(t,z;s,m) = (1 - (1-z)q_2(s,t))^{m} (1 - (1-z)q_1(s,t))^{S-m}. \tag{2.3.29}$$

Here we have for brevity introduced two functions q_1 and q_2 defined as follows:

$$q_1(s,t) = (H(t) - H(s))e^{-\rho(t)}, \tag{2.3.30}$$

$$q_2(s,t) = q_1(s,t) + e^{-(\rho(t)-\rho(s))}. \tag{2.3.31}$$

We show that both of these functions take values in the closed interval from zero to one. Note from (2.3.17) that the derivative of the function ρ equals

$$\rho'(t) = r + h(t)/S. \tag{2.3.32}$$

Since $r > 0$, we get from this relation the inequality

$$h(t) < S\rho'(t). \tag{2.3.33}$$

The definition (2.3.18) of the function $\underline{H}$ shows that

$$H(t) - H(s) = \frac{1}{S} \int_s^t h(u) e^{\rho(u)} du. \tag{2.3.34}$$

By applying the inequality (2.3.33) to this expression, we find that

$$H(t) - H(s) < e^{\rho(t)} - e^{\rho(s)}. \tag{2.3.35}$$

This inequality applied to the definition (2.3.30) of the function q_1 shows that

$$q_1(s,t) < 1 - e^{-(\rho(t)-\rho(s))}. \tag{2.3.36}$$

This inequality for the function $\underline{q}_1$ applied to (2.3.31) shows that

$$q_2(s,t) < 1. \tag{2.3.37}$$

The functions $\underline{q}_1$ and $\underline{q}_2$ are both nonnegative since $\underline{t} \geq \underline{s}$. Hence both take values in the closed unit interval.

The probability generating function $\underline{G}$ is seen from (2.3.29) to be the product of the probability generating functions of two binomially distributed random variables, one with parameters $\underline{m}$ and $\underline{q}_2(\underline{s},\underline{t})$, and the other with parameters $\underline{S}-\underline{m}$ and $q_1(\underline{s},\underline{t})$. This interpretation is allowed since the above finding shows that both $\underline{q}_1$ and $\underline{q}_2$ take values in the unit interval. The probability distribution $P_{mn}(\underline{s},\underline{t})$, where $\underline{n} = 0,1,\ldots,\underline{S}$, is the convolution of these two binomial distributions. The expressions for the transition probability functions are therefore found to be as follows:

$$P_{mn}(s,t) = \sum_{k=0}^{m \wedge n} \binom{m}{k} (q_2(s,t))^k (1 - q_2(s,t))^{m-k}$$

$$\cdot \binom{S-m}{n-k} (q_1(s,t))^{n-k} (1 - q_1(s,t))^{S-m-n+k}. \tag{2.3.38}$$

The expected number of parasites at time $\underline{t}$ is denoted by $\underline{x}(\underline{t})$. For formulation of transmission models we need a differential equation for the function $\underline{x}$. The distribution of $\underline{X}(0)$ is denoted by $\underline{p}_m$:

$$p_m = P\{X(0) = m\}, \quad m = 0,1,\dots,S. \tag{2.3.39}$$

From the expectation for the binomial distribution it follows for arbitrary initial distribution that

$$\begin{aligned} x(t) &= \sum_{m=0}^{S} p_m(mq_2(0,t) + (S-m)q_1(0,t)) \\ &= Sq_1(0,t)\sum_{m=0}^{S} p_m + (q_2(0,t) - q_1(0,t))\sum_{m=1}^{S} mp_m \\ &= Sq_1(0,t) + e^{-\rho(t)}x(0) = (SH(t) + x(0))e^{-\rho(t)}. \end{aligned} \tag{2.3.40}$$

Differentiation of this expression gives the differential equation

$$\begin{aligned} x' &= -\rho'(t)x + SH'(t)e^{-\rho(t)} \\ &= -(r + h(t)/S)x + h(t) \\ &= h(t)(1 - x/S) - rx. \end{aligned} \tag{2.3.41}$$

We investigate the distribution of the number of infections $\underline{X}(\underline{t})$ as $\underline{t} \to \infty$ under the assumption that the immigration rate $\underline{h}(\underline{t})$ approaches a limit as $\underline{t} \to \infty$. Thus, we assume the existence of a nonnegative constant $\underline{H}$, where

$$H = \lim_{t\to\infty} h(t). \tag{2.3.42}$$

It follows then from (2.3.17) and (2.3.18) that

$$\lim_{t\to\infty} (H(t)e^{-\rho(t)}) = H/(H + rS), \tag{2.3.43}$$

while (2.3.30) and (2.3.31) show that

$$\lim_{t\to\infty} q_1(s,t) = H/(H + rS) \tag{2.3.44}$$

and that

$$\lim_{t\to\infty} q_2(s,t) = H/(H + rS). \tag{2.3.45}$$

In this case, therefore, the probability distribution $\underline{P}_{mn}(\underline{s},\underline{t})$, where $\underline{n} = 0,1,\ldots,\underline{S}$, converges to a binomial distribution with parameters $\underline{S}$ and $\underline{H}/(\underline{H} + \underline{rS})$ as $\underline{t} \to \infty$. The same conclusion therefore holds for the convergence of the distribution of $\underline{X}(\underline{t})$, regardless of the initial distribution.

We consider now the distribution of the number of parasites $\underline{X}(\underline{t})$ at time $\underline{t}$ under the assumption that the initial distribution is binomial with parameters $\underline{S}$ and $\underline{p}_0$. It follows then from (2.3.38) for all $t \geq 0$ that

$$\begin{aligned} p_n(t) &= P\{X(t) = n\} \\ &= \sum_{m=0}^{S} \sum_{k=0}^{m} \binom{S}{m}\binom{m}{k}\binom{S-m}{n-k} p_0^m(1 - p_0)^{S-m} \\ &\cdot (q_2(0,t))^k (1 - q_2(0,t))^{m-k} \\ &\cdot (q_1(0,t))^{n-k} (1 - q_1(0,t))^{S-m-n+k} . \end{aligned} \tag{2.3.46}$$

By denoting the general term in this double sum by $\underline{C}_{mk}$, we find that (2.3.46) can be written in the form

$$p_n(t) = \sum_{m=0}^{S} \sum_{k=0}^{m} C_{mk} = \sum_{k=0}^{S} \sum_{m=k}^{S} C_{mk} = \sum_{k=0}^{n} \sum_{j=0}^{S-k} C_{k+j,k}. \tag{2.3.47}$$

We note the binomial identity

$$\binom{S}{m}\binom{m}{k}\binom{S-m}{n-k} = \binom{S}{n}\binom{n}{k}\binom{S-n}{m-k}, \tag{2.3.48}$$

and that

$$C_{k+j,k} = 0 \quad \text{for} \quad j > S-n. \tag{2.3.49}$$

We find therefore that

$$p_n(t) = \binom{S}{n} \sum_{k=0}^{n} \binom{n}{k} (p_0 q_2(0,t))^k ((1-p_0)q_1(0,t))^{n-k}$$

$$\cdot \sum_{j=0}^{S-n} \binom{S-n}{j} (p_0(1 - q_2(0,t)))^j ((1 - p_0)(1 - q_1(0,t)))^{S-n-j}$$

$$= \binom{S}{n} (p_0 q_2(0,t) + (1-p_0)q_1(0,t))^n$$

$$\cdot(p_0(1 - q_2(0,t)) + (1-p_0)(1 - q_1(0,t)))^{S-n}$$

$$= \binom{S}{n} (q_1(0,t) + p_0 e^{-\rho(t)})^n$$

$$\cdot(1 - (q_1(0,t) + p_0 e^{-\rho(t)}))^{S-n} . \qquad (2.3.50)$$

This expression shows that the number of infections $\underline{X}(\underline{t})$ is binomially distributed at each $\underline{t} \geq 0$ with parameters $\underline{S}$ and $\underline{p}(\underline{t}) = \underline{q}_1(0,\underline{t}) + \underline{p}_0 e^{-\rho(t)}$. It is readily shown that $\underline{p}(\underline{t})$ satisfies the differential equation

$$p' = h(t)(1 - p)/S - rp. \qquad (2.4.51)$$

We note that this equation is formally identical with equation (2.1.23) for $\underline{S} = 1$.

We proceed to evaluate the prevalence $\underline{Q}(\underline{s})$, the incidence $\underline{I}(\underline{s},\underline{T})$, and the recovery probability $\underline{R}(\underline{s},\underline{T})$ in the case of constant immigration rate, i.e. with $\underline{h}(\underline{t}) = \underline{H}$. In similarity with the results in Section 2.2, we find the following expressions:

$$Q(s) = 1 - p_0(s), \qquad (2.3.52)$$

$$I(s,T) = 1 - P_{00}(s,s+T), \qquad (2.3.53)$$

$$R(s,T) = 1 - I(s,T) - \frac{Q(s+T) - I(s,T)}{Q(s)} . \qquad (2.3.54)$$

The assumption of constant immigration rate leads to the following expressions for the functions ρ and H:

$$\rho(t) = (r + H/S)t, \qquad (2.3.55)$$

$$H(t) = \frac{H}{H+rS}(e^{(r+H/S)t} - 1). \quad (2.3.56)$$

We define a function β by putting

$$\beta(t) = 1 - e^{-(r+H/S)t}, \quad (2.3.57)$$

and find from (2.3.30) that the function q_1 can be written

$$q_1(s,t) = \frac{H}{H+rS}\beta(t-s). \quad (2.3.58)$$

Expression (2.3.38) shows that the transition probability P_{00} is given by the expression

$$P_{00}(s,t) = (1 - q_1(s,t))^S = (1 - \frac{H}{H+rS}\beta(t-s))^S. \quad (2.3.59)$$

As in the previous section we find that the value of the transition probability function $P_{00}(s,t)$ depends only on the difference, $t-s$, between its time arguments. It follows that both the incidence $I(s,T)$ and the recovery probability $R(s,t)$ can be expressed in terms of values of the function p_0 as follows:

$$I(s,T) = 1 - p_0(T), \quad (2.3.60)$$

$$R(s,T) = \frac{p_0(s+T) - p_0(s)p_0(T)}{1 - p_0(s)}. \quad (2.3.61)$$

Relation (2.2.38) holds also in this case, so we can write

$$p_0(t) = (1 - \frac{H}{H+rS}\beta(t))^S. \quad (2.3.62)$$

It is readily verified that the prevalence increases monotonically with the age toward the value

$$\lim_{s\to\infty} Q(s) = 1 - (rS/(H+rS))^S \quad (2.3.63)$$

Furthermore, we note that the incidence is independent of the age and that it increases monotonically with the length of the time interval T over which it is defined toward the value given by the right-hand side of (2.3.63). The dependence of the recovery rate on $\underline{s}$ and $\underline{T}$ is more complicated.

The host model treated in this section is in a sense more general than the host models treated in the two preceding sections. Indeed, the infection-recovery process of Section 2.1 is recovered from the process of the present section by putting the saturation level $\underline{S}=1$. Furthermore, the superinfection process of Section 2.2 is found from the process of the present section by letting the saturation level $\underline{S}$ approach infinity.

The host model of this section will find an important application in Section 2.7, where we study the number of hosts infected in a population of hosts.

2.4 A Superinfection Process with Monogamous Mating

The sexually mature forms of the schistosome parasites are dioecious, i.e. male and female reproductive organs appear in separate parasites. The infective unit is not the parasite but the mated female parasite. In this section we develop a host model that accounts for the number of mated female parasites in the host. The model can be seen as an extension of the superinfection process without immunity treated in Section 2.2. The hypotheses and notation of that section remain valid. The total number of parasites is modelled by an immigration-death process $\underline{X}(\underline{t})$ with immigration rate $\underline{h}(\underline{t})$ and death rate per parasite $\underline{r}$. Section 2.2 develops several of the important properties of the process $\underline{X}(\underline{t})$.

The number of female and male parasites in the body of the host at time $\underline{t}$ are here modelled by immigration-death processes denoted by $\underline{F}(\underline{t})$ and $\underline{M}(\underline{t})$, respectively. We hypothesize that the immigration rate for each of them is $\underline{h}(\underline{t})/2$ and that the death rate per individual parasite is $\underline{r}$. Thus, any property of $\underline{F}$ or $\underline{M}$ is readily found from the corresponding property of $\underline{X}$ in Section 2.2 by dividing $\underline{h}(\underline{t})$ by 2. In order to be able to deal with joint distributions involving both females and males we assume that the family of random variables $\{\underline{F}(\underline{t})\}$, $t \geq 0$, is independent of the family $\{\underline{M}(\underline{t})\}$, $\underline{t} \geq 0$.

The mating between females and males is hypothesized to be monogamous. Furthermore, it is assumed that pairs of females and males are formed without delay when partners become available. A new immigrant is a potential partner, as is any previously mated parasite after the death of its partner ("widows" and "widowers"). We introduce $\underline{F}_M(\underline{t})$ and $\underline{M}_M(\underline{t})$ to denote the number of mated females and mated males at time $\underline{t}$. Our hypotheses imply that we always have equal numbers of mated females and mated males, and that their common value is equal to the minimum of the number of females and the number of males.

It turns out to be convenient to study also the number of single females, $\underline{F}_S$, and the number of single males, $\underline{M}_S$. The following relations between the random variables introduced are readily seen to hold:

$$X(t) = F(t) + M(t), \tag{2.4.1}$$

$$F_M(t) = M_M(t) = \min(F(t), M(t)), \tag{2.4.2}$$

$$F(t) = F_S(t) + F_M(t), \tag{2.4.3}$$

$$M(t) = M_S(t) + M_M(t). \tag{2.4.4}$$

$$\min(F_S(t), M_S(t)) = 0. \tag{2.4.5}$$

We proceed to derive an expression for the expected number of mated female parasites. This expression is needed in the formulation of a transmission model with the host model of the present section as a building block. Furthermore, we shall derive expressions for the age-dependence of prevalence, incidence and recovery probability in the case of constant immigration rate. For this purpose we define an individual host as being infected whenever his body contains at least one living mated female parasite.

The modified Bessel functions of the first kind, $\underline{I}_k(\underline{x})$ with $\underline{k}$ an integer, play an important role in the distribution of single parasites. For ready reference we quote the definition and some basic properties from Abramovitz-Stegun (1968):

$$I_k(x) = \sum_{m=0}^{\infty} \frac{(x/2)^{2m+k}}{m!(m+k)!} \tag{2.4.6}$$

$$I_k(x) = I_{-k}(x), \tag{2.4.7}$$

$$I_0(x) + 2\sum_{m=1}^{\infty} I_m(x) = e^x, \tag{2.4.8}$$

$$xI_{k-1}(x) - xI_{k+1}(x) = 2kI_k(x), \tag{2.4.9}$$

$$I_{k-1}(x) + I_{k+1}(x) = 2I_k'(x), \tag{2.4.10}$$

$$I_k(x) \sim \frac{e^x}{\sqrt{2\pi x}}, \quad x \to \infty . \tag{2.4.11}$$

By applying the recursion formula (2.4.9) and using relation (2.4.8) we get

$$\begin{aligned} \sum_{k=1}^{\infty} kI_k(x) &= \frac{x}{2}\sum_{k=0}^{\infty} I_k(x) - \frac{x}{2}\sum_{k=2}^{\infty} I_k(x) \\ &= \frac{x}{2}(I_0(x) + I_1(x)) . \end{aligned} \tag{2.4.12}$$

For the following development we make the additional assumption that the initial distributions of $\underline{F}(\underline{t})$ and $\underline{M}(\underline{t})$, i.e. the distributions of $\underline{F}(0)$ and $\underline{M}(0)$, are both Poisson distributions with common parameter $\underline{x}_0/2$. . It follows then from the results in Section 2.2 that $\underline{F}(\underline{t})$ and $\underline{M}(\underline{t})$ both are Poisson-distributed for each $\underline{t} \geq 0$ with time-dependent parameter $\underline{x}(\underline{t})/2$, where the function $\underline{x}$ is given by (2.2.31). The distribution of the number of single females at time $\underline{t}$, $\underline{F}_S(\underline{t})$, is then found as follows. Note first that if $\underline{k}$ is an integer larger than or equal to one, then the number of single females equals $\underline{k}$ exactly when the number of females is equal to the number of males plus $\underline{k}$. Since the number of males can take on any integer value $\underline{m} = 0,1,2,\ldots$, we find that

$$\{F_S(t)=k\} = \bigcup_{m=0}^{\infty}\{F(t) = m+k, \quad M(t) = m\} \ , \ k=1,2,\dots \ . \qquad (2.4.13)$$

By using the independence of $\underline{F}(\underline{t})$ and $\underline{M}(\underline{t})$ and the fact that both have Poisson distributions with parameter $\underline{x}(\underline{t})/2$ we get

$$P\{F_S(t)=k\} = \sum_{m=0}^{\infty} \frac{(x(t)/2)^{m+k}}{(m+k)!} e^{-x(t)/2} \frac{(x(t)/2)^m}{m!} e^{-x(t)/2}$$

$$= e^{-x(t)} I_k(x(t)), \quad k = 1,2,\dots \ . \qquad (2.4.14)$$

Here we have used the definition of $\underline{I}_k$ in (2.4.6). The probability that the number of single females is positive is therefore

$$P\{F_S(t) > 0\} = \sum_{k=1}^{\infty} P\{F_S(t) = k\} = e^{-x(t)} \sum_{k=1}^{\infty} I_k(x(t))$$

$$= \frac{1 - e^{-x(t)} I_0(x(t))}{2} , \qquad (2.4.15)$$

which we find by an application of (2.4.8). Note next that the number of single females equals zero if either there is a positive number of single males, or if both the number of single females and the number of single males is zero. The latter event occurs when the number of females is equal to the number of males. Thus we have

$$\{F_S(t)=0\} = \{M_S(t) > 0\} \cup \bigcup_{f=0}^{\infty} \{F(t) = f, M(t)=f\} , \qquad (2.4.16)$$

and hence

$$P\{F_S(t) = 0\} = \frac{1 - e^{-x(t)} I_0(x(t))}{2} + \sum_{f=0}^{\infty} \frac{(x(t)/2)^{2f}}{f!f!} e^{-x(t)}$$

$$= \frac{1}{2} + \frac{1}{2} e^{-x(t)} I_0(x(t)). \qquad (2.4.17)$$

Here we have used the fact that the number of single males has the same distribution as the number of single females.

The expected number of single females at time $\underline{t}$ is from (2.4.14) and (2.4.12) found to be

$$EF_S(t) = \sum_{k=1}^{\infty} k\ e^{-x(t)} I_k(x(t)) =$$

$$= \frac{x(t)}{2}\ e^{-x(t)}\ (I_0(x(t)) + I_1(x(t))). \qquad (2.4.18)$$

The total population is clearly composed of single and mated females and males:

$$X(t) = F_S(t) + F_M(t) + M_S(t) + M_M(t). \qquad (2.4.19)$$

In this relation there is equality between the number of mated females and the number of mated males. Furthermore, the number of single males has the same expectation as the number of single females. We find therefore that the expected number of mated females can be written as follows:

$$EF_M(t) = \frac{1}{2}\ EX(t) - EF_S(t)$$

$$= \frac{1}{2}\ x(t) - \frac{1}{2}\ x(t)e^{-x(t)}(I_0(x(t)) + I_1(x(t)))$$

$$= \frac{1}{2}\ \Psi(x(t)), \qquad (2.4.20)$$

where the function Ψ is defined by

$$\Psi(x) = x(1 - e^{-x}(I_0(x) + I_1(x))). \qquad (2.4.21)$$

The function Ψ gives the expected number of mated parasites (female + male) as a function of the total expected number of parasites. We proceed to give some properties of the function Ψ. By differentiation and application of (2.4.10), (2.4.7) and (2.4.9) we find that

$$\Psi'(x) = 1 - e^{-x}I_0(x) \qquad (2.4.22)$$

and that

$$\Psi''(x) = e^{-x}(I_0(x) - I_1(x)). \qquad (2.4.23)$$

For $\underline{x} > 0$ we have $\underline{I}_k(\underline{x}) > 0$ and hence, using (2.4.8),

$$0 < \Psi(x) < x, \quad x > 0, \tag{2.4.24}$$

$$\Psi'(x) > 0, \quad x > 0. \tag{2.4.25}$$

From Soni (1965) it follows that

$$I_{\nu+1}(x) < I_\nu(x), \quad x > 0, \quad \nu > -1/2. \tag{2.4.26}$$

Hence we find from (2.4.23) that

$$\Psi''(x) > 0, \quad x > 0. \tag{2.4.27}$$

From the series expansion (2.4.6) we get

$$I_k(x) = \frac{1}{k!}(x/2)^k + 0(x^{k+2}), \quad x \to 0, \tag{2.4.28}$$

and hence

$$\Psi(x) \sim \frac{1}{2} x^2, \quad x \to 0. \tag{2.4.29}$$

Finally, (2.4.11) gives

$$\Psi(x) \sim x - \sqrt{2x/\pi}, \quad x \to \infty. \tag{2.4.30}$$

We proceed to derive expressions for the prevalence $\underline{Q}_d(\underline{s})$, the incidence $\underline{I}_d(\underline{s},\underline{T})$, and the recovery probability $\underline{R}_d(\underline{s},\underline{t})$. The subscript $\underline{d}$ reminds us that we are dealing with dioecious parasites and that an individual host is infected whenever his body contains at least one mated female parasite. The prevalence, incidence, and recovery probability are all defined in terms of the distribution and the transition probabilities of $\underline{F}_M$, the number of mated females. The first step is to express them in terms of the corresponding properties of $\underline{F}$, the total number of females. For this step we make use of the relation

$$\{F_M(s) > 0\} = \{F(s) > 0, \; M(s) > 0\}, \tag{2.4.31}$$

which merely says that the number of mated females is positive when there is at least one female and one male present. We note that this identity between events holds for any form of mating between females and males and not only for the monogamous mating assumed in this section.

For the prevalence at age $\underline{s}$ we find

$$Q_d(s) = P\{F(s) > 0,\ M(s) > 0\} = (P\{F(s) > 0\})^2$$

$$= (1 - p_{F0}(s))^2. \qquad (2.4.32)$$

Here we have used the facts that $\underline{F}(\underline{s})$ and $\underline{M}(\underline{s})$ are independent and have identical distributions. In the last equality in (2.4.32) we have introduced $\underline{p}_{Fk}(\underline{s})$ to denote the distribution of $\underline{F}(\underline{s})$, i.e.

$$p_{Fk}(s) = P\{F(s) = k\}, \quad k = 0,1,\ldots \;. \qquad (2.4.33)$$

The recovery probability at age $\underline{s}$ over a time interval of length $\underline{T}$ can be written

$$R_d(s,T) = 1 - P\{F_M(s+T) > 0 | F_M(s) > 0\}. \qquad (2.4.34)$$

We evaluate the transition probability on the right-hand side with the aid of (2.4.31). In the evaluation we use the independence of $\{\underline{F}(\underline{t})\}$, $\underline{t} \geq 0$, and $\{\underline{M}(\underline{t})\}$, $\underline{t} \geq 0$. We get

$$1 - R_d(s,T) = \frac{1}{P\{F_M(s)>0\}} \cdot P\{F(s+T)>0,\ M(s+T)>0,\ F(s)>0,\ M(s)>0\}$$

$$= \frac{1}{P\{F_M(s)>0\}} \cdot (P\{F(s+T) > 0,\ F(s) > 0\})^2$$

$$= (P\{F(s+T) > 0 | F(s) > 0\})^2. \qquad (2.4.35)$$

By using the expression (2.2.42) for the transition probability $\underline{P}\{\underline{F}(\underline{s}+\underline{T}) = 0 | \underline{F}(\underline{s}) > 0\}$ we get

$$R_d(s,T) = 1 - \left(1 - \frac{p_{F0}(s+T) - p_{F0}(s)p_{F0}(T)}{1 - p_{F0}(s)}\right)^2 . \qquad (2.4.36)$$

To derive an expression for the incidence we make use of the following general relation between prevalence, incidence, and recovery probability:

$$Q_d(s+T) = Q_d(s)(1-R_d(s,T)) + (1-Q_d(s))I_d(s,T). \quad (2.4.37)$$

The corresponding relation in the monoecious case was established in (2.2.36) and was used there to derive the expression already used above for the recovery probability.

We note from (2.4.32) that

$$1 - Q_d(s) = 2p_{FO}(s) - p_{FO}^2(s) . \quad (2.4.38)$$

By solving (2.4.37) for the incidence $\underline{I}_d(\underline{s},\underline{T})$ and using the expressions derived above for the prevalence and the recovery probability we get

$$I_d(s,T) = \frac{1}{2p_{FO}(s) - p_{FO}^2(s)} \left[(1 - p_{FO}(s+T))^2 - (1 - p_{FO}(s))^2 \left(1 - \frac{p_{FO}(s+T) - p_{FO}(s)p_{FO}(T)}{1 - p_{FO}(s)}\right)^2 \right]$$

$$= (1 - p_{FO}(T)) \left(1 - \frac{2p_{FO}(s+T) - p_{FO}(s)p_{FO}(T)}{2 - p_{FO}(s)}\right). \quad (2.4.39)$$

Reference to (2.2.43) shows that the function $\underline{p}_{FO}$ can be written as follows:

$$p_{FO}(t) = \exp(- \frac{H}{2r} \alpha(t)) \quad , \quad (2.4.40)$$

where the function α is defined in (2.2.44).

By using (2.4.40) in the expressions for the prevalence $\underline{Q}_d(\underline{s})$ in (2.4.32), the incidence $\underline{I}_d(\underline{s},\underline{T})$ in (2.4.39), and the recovery probability $\underline{R}_d(\underline{s},\underline{T})$ in (2.4.36), we get explicit expressions for each of these three epidemiological quantities in terms of the infection rate $\underline{H}$, the death rate per parasite $\underline{r}$, the age of the host $\underline{s}$, and the length $\underline{T}$ of the time interval over which incidence and recovery probability are defined.

The prevalence $\underline{Q}_d(\underline{s})$ is seen to increase monotonically with the host age $\underline{s}$ toward the value $(1 - \exp(\underline{H}/2\underline{r}))^2$. The incidence $\underline{I}_d(\underline{s},\underline{T})$ is a monotonically increasing function of both the age $\underline{s}$ and the length $\underline{T}$ of the time interval over which it is defined. The recovery probability $\underline{R}_d(\underline{s},\underline{T})$ decreases monotonically with the host age $\underline{s}$, while its dependence on $\underline{T}$ is more complicated. Details are given in Appendix II.

2.5 A Superinfection Process with Polygamous Mating

Our knowledge about the mating pattern of schistosomes is incomplete and direct observation of live schistosomes in blood vessels of human beings is practically impossible. This implies that it is difficult to substantiate the hypothesis of monogamous mating between males and females. We proceed therefore in this section to develop a host model based on the alternative hypothesis that mating between male and female parasites is polygamous. Specifically, we assume that one male suffices to fertilize all females in the body of a definitive host. Thus the number of mated females equals the total number of females if at least one male is present, but is zero otherwise.

As in the preceding section we assume that the distributions of $\underline{F}(0)$ and $\underline{M}(0)$ are Poisson with common parameter $\underline{x}_0/2$. We shall use the noteworthy consequence of this assumption that both $\underline{F}(\underline{t})$ and $\underline{M}(\underline{t})$ are Poisson-distributed for each $\underline{t} \geq 0$ with timedependent parameter $\underline{x}(\underline{t})/2$, where the function $\underline{x}$ is given by (2.2.31). The main result to be derived is an expression for the expected number of mated females.

Our mating hypothesis leads us to conclude that the number of mated females is equal to $\underline{k} \geq 1$ when the total number of females equals $\underline{k}$ and there is at least one male present. Thus

$$\{F_M(t) = k\} = \{F(t) = k,\ M(t) > 0\}, \quad k = 1,2,\ldots \tag{2.5.1}$$

By using the independence of $\underline{F}(\underline{t})$ and $\underline{M}(\underline{t})$ we find that

$$P\{F_M(t) = k\} = P\{F(t) = k\} \cdot P\{M(t) > 0\}\ , \ k = 1,2,\ldots \tag{2.5.2}$$

From this expression we find that the expected number of mated females is equal to

$$EF_M(t) = P\{M(t) > 0\} \cdot EF(t). \tag{2.5.3}$$

By finally using the distributional properties of $\underline{F}(\underline{t})$ and $\underline{M}(\underline{t})$ we conclude that

$$EF_M(t) = \frac{1}{2}\,(1 - e^{-x(t)/2})x(t). \tag{2.5.4}$$

This expression is needed for establishing a hybrid transmission model that uses the present host model as a building block.

The expressions for prevalence, incidence and recovery probability of the preceding section hold also for the host model of the present section since they are based on relation (2.4.31), which holds for both monogamous and polygamous mating.

2.6 Infection in a Population of Hosts

The host models considered so far are all established with reference to one individual host. The essential state variable in these models corresponds to the number of infections or the number of parasites in the host. We proceed to establish two models that account for the number of infected hosts in a population of hosts. The infection of each host is modelled by the simple infection-recovery process treated in Section 2.1. There are two states for each host. Either the host is uninfected and susceptible to infection, or he is infected and immune to additional infections. We assume homogeneity of all hosts with respect to both exposure and susceptibility. This means that the infection rate $h(t)$ per uninfected host is the same for all hosts.

In the first model, the population of hosts is assumed constant, and the number of hosts is denoted by N. Infected hosts are assumed to recover with the constant rate r per infected host. We introduce $X(t)$, $t \geq 0$, to denote the number of infected hosts at time t. The possible values of $X(t)$ are $0,1,2,\ldots,N$. The transition probabilities for $X(t)$ are denoted $P_{mn}(s,t)$. The hypotheses of the model take the following form:

$$P_{n,n+1}(t,t+\Delta t) = (N-n)h(t)\Delta t + o(\Delta t), \quad n=0,1,\ldots,N-1, \tag{2.6.1}$$

$$P_{n,n-1}(t,t+\Delta t) = nr\Delta t + o(\Delta t), \quad n=1,2,\ldots,N, \tag{2.6.2}$$

$$P_{n,n}(t,t+\Delta t) = 1 - ((N-n)h(t) + nr)\Delta t + o(\Delta t). \tag{2.6.3}$$

Thus, the probability for the number of infected hosts to increase by one during a short time interval is essentially equal to the product of the infection rate per uninfected host, the number of uninfected hosts, and the length of the time interval. Similarly, the probability for the number of infected hosts to decrease by one during a short time interval is essentially equal to the product of the recovery rate per infected host, the number of infected hosts, and the length of the time interval.

The inhomogeneous Markov chain whose transition probabilities satisfy the above hypotheses has already been analyzed. A comparison of (2.6.1) - (2.6.3) with the hypotheses (2.3.1) - (2.3.3) in the superinfection process with immunity studied in Section 2.3 shows that the two models are essentially equivalent. All we need to do in order to describe the various properties of $X(t)$ in the present section is to use the corresponding results in Section 2.3 and replace the saturation level S by the host population size N and furthermore replace the infection rate $h(t)$ by $Nh(t)$. Thus we find from (2.3.41) that the expected number of infected hosts $x(t)$ satisfies the differential equation

$$x' = h(t)(N - x) - rx. \tag{2.6.4}$$

The distribution of $\underline{X}(\underline{t})$ is particularly simple if the initial distribution, i.e. the distribution of $\underline{X}(0)$, is binomial with parameters $\underline{N}$ and $\underline{p}_0$. Indeed, relation (2.3.50) shows that $\underline{X}(\underline{t})$ is then binomially distributed for all $\underline{t} \geq 0$, with parameters $\underline{N}$ and $\underline{p}(\underline{t})$, where $\underline{p}(\underline{t})$ satisfies the differential equation

$$p' = h(t)(1-p) - rp. \tag{2.6.5}$$

This equation is identical to equation (2.1.23) for the infection probability of an individual host. The solutions of the two equations are equal if the initial conditions coincide. We proceed to show that the assumption that $\underline{X}(0)$ has a binomial distribution with parameters $\underline{N}$ and $\underline{p}_0$ implies that the function $\underline{p}(\underline{t})$ for the model treated here is equal to each of the infection probabilities $\underline{p}_i(\underline{t})$ for the individual hosts $\underline{i} = 1,2,\ldots,\underline{N}$.

To prove this remark we note first that the random variable $\underline{X}(\underline{t})$ can be written as the sum of $\underline{N}$ independent random variables $\underline{X}^{(i)}(\underline{t})$, $\underline{i} = 1,2,\ldots,\underline{N}$:

$$X(t) = X^{(1)}(t) + \ldots + X^{(N)}(t). \tag{2.6.6}$$

Here, each $\underline{X}^{(i)}(\underline{t})$ takes the value one if host $\underline{i}$ is infected at time $\underline{t}$, and is equal to zero otherwise. Let the initial probability that host $\underline{i}$ is infected be given by

$$p_i = P\{X^{(i)}(0)\}. \tag{2.6.7}$$

Equality between means and between variances of left- and right-hand sides of (2.6.6) at $\underline{t} = 0$ gives

$$Np_0 = \sum_{i=1}^{N} p_i \tag{2.6.8}$$

and

$$Np_0(1-p_0) = \sum_{i=1}^{N} p_i(1-p_i). \tag{2.6.9}$$

These two equations can be used to solve for the $\underline{N}$ unknowns $\underline{p}_i$, $\underline{i} = 1,2,\ldots,\underline{N}$, in terms of $\underline{p}_0$. We get

$$p_0 = \frac{1}{N} \sum_{i=1}^{N} p_i, \tag{2.6.10}$$

and

$$p_0^2 = \frac{1}{N} \sum_{i=1}^{N} p_i^2. \tag{2.6.11}$$

By using (2.6.10), the latter relation can be written

$$p_0^2 = \frac{1}{N} \sum_{i=1}^{N} (p_i - p_0)^2 + p_0^2. \tag{2.6.12}$$

This implies that the only solution of these equations is

$$p_i = p_0, \quad i = 1,2,\ldots,N. \tag{2.7.12}$$

It follows that all the initial infection probabilities $p_i(0)$ associated with the individual hosts i are equal to the initial value p_0 for the model treated here. We conclude that $p(t) = p_i(t)$, $t \geq 0$, $i = 1,2,\ldots,N$.

The second model to be studied concerns the situation where the population size is constant and equal to N, but where individual hosts are subject to birth and death. The death rate per host is denoted by r. The hosts are subject to infection with the infection rate $h(t)$ per uninfected host. Infected hosts are not assumed to recover from the infection; they remain infected until they die.

In order for the population size to remain constant, we assume that each time that a host dies, it is replaced by a newborn one. Newborn hosts are assumed to be uninfected. This has the consequence that death of an uninfected host does not change the number of uninfected or infected hosts, while death of an infected hosts leads to a decrease of the number of infected hosts by one, and a simultaneous increase of the number of uninfected hosts by one. It is readily seen that this transition corresponds to recovery of the associated host in the model treated above. Indeed, all the results discussed for the first model hold also for the second one when we are concerned only with the distribution of the number of infected hosts $X(t)$.

It is important to note that the expressions for prevalence, incidence, and recovery probability of the present models are those developed in Section 2.1, and not those of Section 2.3, since these concepts describe the infectious status of individual hosts.

2.7. Host Models with Latency

The transmission of any infectious disease is affected by latency. The latent period is defined as the time lapse between the moment when an individual receives infectious material and the moment when it is first possible to ascertain that the infectious material has developed in the body of the host. The host is then called infected. Whether or not the host is also infective, i.e. in a position to spread the infection to others, may depend on additional circumstances. For indirectly transmitted infections there are latent periods in both the host populations. Many mathematical models for the transmission of parasitic infections ignore the latent periods. This amounts to an approximation that may be quite acceptable when the latent period is short compared to the period when the host is infected. This requirement is often fulfilled with regard to the human population. However, for the vector population (mosquitoes for malaria, snails for schistosomiasis and hermaphroditic helminthiasis), the latent period is often of the same order of magnitude as the life length of the host. The latent period can then have an appreciable influence on the transmission of the infection, since many mosquitoes or snails die after receiving infectious material but before becoming infected. In order to assess this influence we need to establish transmission models that account for the latent period. As a preparation for this we study host models with latency in the present section. These host models will be used as building blocks for transmission models in Chapter 3.

The models we deal with here are deterministic. The methodology of hybrid models that is used in Chapter 3 allows transmission models to be established on the basis of two host models of which one is stochastic and the other one is deterministic. Three different models will be considered.

For the first model we introduce three state variables S, L, and I to denote the number of hosts that are susceptible, in latent state, and infected, respectively. We assume that the population of hosts is subject to immigration and death, with an immigration rate A and a death rate r per host. We furthermore assume that the susceptible hosts at time t are exposed to infection with an infection rate $h(t)$ per susceptible host. We shall not consider variation in the number of hosts. We assume therefore that the initial population size is N, i.e. $S(0) + L(0) + I(0) = N$, and that $A = rN$. These assumptions imply that the number of hosts is equal to N for each $t \geq 0$. We introduce k to denote the rate at which latent hosts become infected. Our hypotheses lead to the following system of differential equations for S, L, I:

$$S' = rN - h(t)S - rS, \qquad (2.7.1)$$

$$L' = h(t)S - kL - rL, \qquad (2.7.2)$$

$$I' = kL - rI. \qquad (2.7.3)$$

Under the assumption that $h(t)$ approaches the constant value H as $t \to \infty$, we find the following equilibrium values for S, L, I:

$$\bar{S} = \frac{r}{H+r} N, \qquad (2.7.4)$$

$$\bar{L} = \frac{r}{k+r} \frac{H}{H+r} N, \qquad (2.7.5)$$

$$\bar{I} = \frac{k}{k+r} \frac{H}{H+r} N. \qquad (2.7.6)$$

The first factor in the right-hand side of (2.7.6) shows that latency reduces the steady-state number of infected hosts by the factor $k/(k + r)$, which is the proportion of hosts that survive the latent period in steady state.

The assumption that the rate $\underline{k}$ at which latent hosts become infected is constant would in a stochastic formulation of the model imply that the time spent in the latent state has an exponential distribution. It may in many cases be more realistic to assume that the time spent in the latent state is constant.

The second model is based on this assumption. We introduce $\underline{T}$ to denote the constant time spent in the latent state. The hypotheses lead to the following system of functional differential equations for the functions $\underline{S}$, $\underline{L}$, $\underline{I}$:

$$S' = rN - h(t)S - rS, \tag{2.7.7}$$

$$L' = h(t)S - e^{-rT}h(t-T)S(t-T) - rL, \tag{2.7.8}$$

$$I' = e^{-rT}h(t-T)S(t-T) - rI. \tag{2.7.9}$$

The arguments of the state variables are deleted when they are equal $\underline{t}$ but indicated otherwise. The rate at which latent hosts become infected at time $\underline{t}$ is equal to the rate at which susceptible hosts became latent $\underline{T}$ units of time earlier, multiplied by the proportion $\underline{e}^{-rT}$ of hosts that survive the latent period. If this system of equations is to be solved for $\underline{t} \geq 0$, we require that the initial values $\underline{L}(0)$ and $\underline{I}(0)$ be given, and that in addition the values of $\underline{h}(\underline{t})$ and $\underline{S}(\underline{t})$ be given over the initial time interval $[-\underline{T}, 0]$, where $0 \leq \underline{S}(\underline{t}) \leq \underline{N}$. We assume as above that $\underline{S}(0) + \underline{L}(0) + \underline{I}(0) = \underline{N}$, so that the total population size remains constant. If $\underline{h}(\underline{t}) \to \underline{H}$ as $\underline{t} \to \infty$, then we find the following equilibrium values:

$$\bar{S} = \frac{r}{H+r} N, \tag{2.7.10}$$

$$\bar{L} = (1 - e^{-rT}) \frac{H}{H+r} N, \tag{2.7.11}$$

$$\bar{I} = e^{-rT} \frac{H}{H+r} N. \tag{2.7.12}$$

A comparison between (2.7.6) and (2.7.12) shows that latency in both models reduces the steady - state number of infected hosts by a factor which equals the proportion of hosts that survive the latent period in steady state. The magnitude of the reduction is different, but the qualitative effect is the same.

The third model to be considered is more realistic: it allows for differential mortality and for recovery of infected hosts. An additional state variable is introduced, namely the number of recovered hosts $R(t)$. Recovered hosts are assumed not to be susceptible to additional infections. Different death rates are allowed for the different categories of hosts. The death rates are denoted r_S, r_L, r_I, and r_R, where the subscript indicates the host category that each death rate applies to. It is natural to assume that the death rate is smallest for susceptible hosts and that therefore the following inequality holds:

$$r_S \leq \min(r_L, r_I, r_R). \tag{2.7.13}$$

It is also natural to assume that the death rate is largest for infected hosts. The differential mortality would be degenerate if all the death rates were equal. In order to avoid this situation we assume that

$$r_S < \max(r_L, r_I, r_R) = r_I. \tag{2.7.14}$$

This inequality combined with a constant rate of immigration of susceptible hosts implies that the total number of hosts becomes smaller if the infection rate $h(t)$ is increased. We introduce N to denote the maximum steady-state number of hosts, which is achieved when the infection rate is identically equal to zero. In that case all hosts are susceptible in the steady state. The rate of immigration of susceptible hosts is assumed to be constant, independent of the infection rate. It is therefore equal to $r_S N$. The rate at which latent hosts become infected is denoted k_L, and the rate at which infected hosts recover is denoted k_I. These hypotheses lead to the following system of differential equations for the four state variables S, L, I, R:

$$S' = r_S N - h(t)S - r_S S \; , \tag{2.7.15}$$

$$L' = h(t)S - k_L L - r_L L \; , \tag{2.7.16}$$

$$I' = k_L L - k_I I - r_I I , \tag{2.7.17}$$

$$R' = k_I I - r_R R . \tag{2.7.18}$$

If $\underline{h}(\underline{t}) \to \underline{H}$ as $\underline{t} \to \infty$, then the solution of this linear system of equations approaches an equilibrium value with the following components:

$$\bar{S} = \frac{r_S}{H + r_S} N , \tag{2.7.19}$$

$$\bar{L} = \frac{r_S}{k_L + r_L} \frac{H}{H + r_S} N , \tag{2.7.20}$$

$$\bar{I} = \frac{r_S}{k_I + r_I} \frac{k_L}{k_L + r_L} \frac{H}{H + r_S} N , \tag{2.7.21}$$

$$\bar{R} = \frac{r_S}{r_R} \frac{k_I}{k_I + r_I} \frac{k_L}{k_L + r_L} \frac{H}{H + r_S} N. \tag{2.7.22}$$

Expression (2.7.21) shows that latency, differential mortality and recovery combine to reduce the steady-state number of infected hosts by the product of two factors. One of them, $\underline{k}_L/(\underline{k}_L + \underline{r}_L)$, measures, as above, the proportion of those hosts that have become latent that survive the latent period. The other reducing factor, $\underline{r}_S/(\underline{k}_I + \underline{r}_I)$, measures the relative reduction due to differential mortality and recovery of the expected time during which a host remains infected.

The ideas illustrated above can easily be used to modify this model in the sense of replacing the assumption of a constant rate for latent hosts to become infected by the assumption that the time spent in the latent state is constant.

CHAPTER 3. TRANSMISSION MODELS FOR MALARIA

We are concerned in this and the following two chapters with the formulation and study of transmission models for malaria, schistosomiasis, and hermaphroditic helminthiasis. Any transmission model can be viewed as a model for the population of parasites in an ecological community. The size of the parasite population is an indication of the severity of infection in the community. A basic purpose of a transmission model is to find how the size of the parasite population is determined by biological and environmental influences. Once this purpose has been achieved, the model can also be used to evaluate the efforts required for eradication and the efficiencies of various modes of control.

The diseases dealt with have the property in common that the causative parasite spends one phase of its life cycle in a human host and another phase in an alternative host (mosquito for malaria, snail for schistosomiasis and hermaphroditic helminthiasis). We shall use host models from Chapter 2 for each of these two phases. The corresponding hybrid transmission model is formed by a deterministic coupling between the stochastic host models or between one stochastic and one deterministic host model. The hypotheses of each transmission model are based on a set of parameters whose biological relevance will be carefully described.

3.1 The Ross Malaria Model

The building of realistic models for the transmission of malaria is a great and important challenge in mathematical epidemiology. The first work in this area is that of Sir Ronald Ross (1909, 1911). The Ross model is deterministic and reflects the basic mechanism that both human beings and mosquitoes are necessary for the transmission of the infection. It has been used to establish an important threshold result and to study the effects of various methods of controlling a malaria infection. A description of the model is given by Bailey (1982).

We proceed to formulate and develop a hybrid model corresponding to the hypotheses of the Ross model. The Ross model is not sufficiently detailed to be accepted as a realistic model for the transmission of malaria. Still, it is thought worthwhile to give a detailed discussion of the hybrid version of the Ross model. The main purpose is to present our methodology in as simple a setting as possible. The various results that will be derived can be expected to have counterparts in more realistic and hence more complex models.

3.1.1 Model Formulation

Before the hypotheses can be described, we introduce seven parameters that measure the biological and environmental influences on the parasite population. Parameters that refer to the human population carry the subscript 1, while parameters that refer to the mosquito population carry the subscript 2. The basic parameters are as follows:

$\underline{N}_1$ = the number of human hosts (assumed constant),

$\underline{r}_1$ = the rate of recovery of an infected human host,

$\underline{b}_1$ = the infectivity of an infected human host, defined as the probability that a bite by a susceptible mosquito on an infected human being will transfer infection to the mosquito,

$\underline{N}_2$ = the number of (female) mosquitoes (assumed constant),

$\underline{r}_2$ = the death rate per mosquito,

$\underline{b}_2$ = the infectivity of an infected mosquito, defined as the probability that a bite by an infected mosquito on a susceptible being will transfer infection to the human being,

$\underline{a}_2$ = the man-biting rate of the mosquitos, defined as the average number of bites on human beings by each (female) mosquito in unit time.

We note that the number of mosquitoes is assumed to be constant and that the mosquitoes are assumed to die with a constant death rate. For consistency between these two assumptions we assume as in Section 2.6 that each time a mosquito dies it is replaced by a newborn one.

Each human host and each mosquito host is assumed to be either uninfected and susceptible to infection or infected and immune to additional infections. We further assume homogeneity of all human hosts and of all mosquito hosts with respect to exposure, susceptibility, and recovery.

Our mathematical model consists of two interdependent Markov chains $X_1(t)$ and $X_2(t)$. Here, $X_1(t)$ can be interpreted as the number of infected human beings at time t, and $X_2(t)$ can be interpreted as the number of infected mosquitoes at time t. The state space of $X_1(t)$ is $\{0,1,2,\ldots,N_1\}$, and the state space of $X_2(t)$ is $\{0,1,2,\ldots,N_2\}$. Infection of an uninfected human being at time t corresponds to an increase by one of the state of the Markov chain $X_1(t)$, while recovery of an infected host corresponds to a decrease by one. The infection rate per uninfected human being is denoted by $h_1(t)$, while the recovery rate per infected human being is the parameter r_1 introduced above. Similarly, infection of an uninfected mosquito at time t corresponds to an increase by one of the state of the Markov chain $X_2(t)$. We denote the infection rate per uninfected mosquito by $h_2(t)$. Death of a mosquito is by our hypotheses accompanied by the birth of an uninfected mosquito. We identify the newborn mosquito with the dead one that it replaces. In this way, the death of an uninfected mosquito leads to no state change in the Markov chain $X_2(t)$. The death of an infected mosquito at time t corresponds to a decrease by one of the state of the Markov chain $X_2(t)$. The rate per infected mosquito with which such transitions take place is the parameter r_2 introduced above.

Both versions of the model for infection in a population of hosts dealt with in Section 2.6 will be used. The first one, which deals with infection and recovery in a constant population of hosts, applies to X_1, while the second one, which deals with infection in a population of constant size subject to birth and death, applies to $\underline{X}_2$.

We assume that the initial distributions of the two random variables $\underline{X}_1(0)$ and $\underline{X}_2(0)$ are both binomial. Specifically, the assumptions are that $\underline{X}_1(0)$ has a binomial distribution with parameters $\underline{N}_1$ and $\underline{p}_{10}$ and that $\underline{X}_2(0)$ has a binomial distribution with parameters $\underline{N}_2$ and $\underline{p}_{20}$. It follows then from the arguments in Section 2.6 that the infection probabilities at each $\underline{t} \geq 0$ are the same for all human hosts and also the same for all mosquito hosts. We denote them by $\underline{p}_1(\underline{t})$ and $\underline{p}_2(\underline{t})$, respectively. It follows from (2.6.5) that they satisfy the following system of differential equations:

$$p_1' = h_1(t)(1-p_1) - r_1 p_1, \tag{3.1.1}$$

$$p_2' = h_2(t)(1-p_2) - r_2 p_2. \tag{3.1.2}$$

The model specification is not yet complete, since the infection rates $\underline{h}_1(\underline{t})$ and $\underline{h}_2(\underline{t})$ have not yet been expressed in terms of state variables and parameters. In order to do this, we explore the consequences of the assumption of a constant man-biting rate. The man-biting rate $\underline{a}_2$ of each mosquito is interpreted to mean that the probability for a given mosquito to bite any human being during a time interval of length $\Delta\underline{t}$ is $\underline{a}_2\Delta\underline{t} + o(\Delta\underline{t})$, $\Delta\underline{t} \to 0$. The assumption of uniform exposure leads us to conclude that the rate at which a given mosquito bites a given human being is $\underline{a}_2/\underline{N}_1$. From this it follows that the rate $\underline{h}_1(\underline{t})$ at which a given human being becomes infected at time $\underline{t}$ is equal to $\underline{a}_2/\underline{N}_1$ multiplied by the number of infective mosquitoes at time $\underline{t}$. This number is the product of the infectivity $\underline{b}_2$ of the mosquitoes and the number of infected mosquitoes. The latter number is a stochastically varying quantity. One of the two hybrid assumptions of the model involves replacing this stochastic quantity by its expected value $\underline{N}_2\underline{p}_2(\underline{t})$. Thus, we get

$$h_1(t) = a_2 b_2 N_2 p_2(t)/N_1 \tag{3.1.3}$$

In a similar way we find that the rate $\underline{h}_2(\underline{t})$ at which a given mosquito becomes infected at time $\underline{t}$ is equal to $\underline{a}_2/\underline{N}_1$ multiplied by the number of infective human beings at time t. This number is the product of the infectivity b_1 of the human beings and the number of infected human beings. The second hybrid assumption of the model involves replacing the stochastically varying number of infected human beings by its expected value $N_1 p_1(t)$. Thus we get

$$h_2(t) = a_2 b_1 p_1(t) \tag{3.1.4}$$

By inserting these expressions for the infection rates into the equations (3.1.1), (3.1.2), we are led to the following system of differential equations for the infection probabilities $p_1(t)$ and $p_2(t)$:

$$p_1' = \frac{a_2 b_2 N_2}{N_1} p_2(1-p_1) - r_1 p_1, \tag{3.1.5}$$

$$p_2' = a_2 b_1 p_1(1-p_2) - r_2 p_2. \tag{3.1.6}$$

Initial values for this system of equations are specified at $t = 0$ by p_{10} and p_{20}. The initial point (p_{10}, p_{20}) belongs to the unit square

$$S = \{(p_1, p_2) : 0 \leq p_1 \leq 1, \ \ 0 \leq p_2 \leq 1\} . \tag{3.1.7}$$

Nonlinearities appear in both of the equations (3.1.5) and (3.1.6). This means that the Ross malaria model has density dependent mechanisms both in the phase of the parasite life cycle spent in human hosts in and in the phase spent in mosquitoes. The density dependencies are caused by the strong immune reaction of the host model of Section 2.1.

3.1.2 Quasi-Dimensional Analysis

Before we proceed with a mathematical analysis of the system of differential equations (3.1.5), (3.1.6), we shall carry through a so-called quasi-dimensional analysis. This can be seen as an extension of the technique of dimensional analysis that has been applied with great success on a variety of physical problems, as discussed by Lin and Segel (1974). Our discussion in this subsection is more elaborate that what is motivated by the rather simple equations in (3.1.5), (3.1.6). It aims to give a background for quasi-dimensional analysis of all the systems of equations that appear in the study of transmission models in later sections.

The basic idea in dimensional analysis is to replace all variables appearing in the equations to be analyzed by dimensionless quantities. One of the great advantages with dimensional analysis is that it leads to a reduction of the number of parameters that determine the solution. This does not mean that any of the basic parameters disappear, but rather that they always appear in certain combinations that can be used to define new parameters that are fewer than the original ones.

Both parameters and variables in physical problems can be given dimensions in terms of fundamental units such as mass, length, and time. Dimensionless variables have the property that their numerical values are the same whatever unit of measurement is used for the fundamental units. In other words, dimensional analysis removes the dependence on arbitrary units of measurement. The only physical dimension in our system of equations (3.1.5), (3.1.6) is that of time. In addition, there are parameters and variables that measure or depend on numbers of human beings, numbers of mosquitoes, and numbers of bites by mosquitoes. For these there is no arbitrariness in the unit of measurement. The natural unit is one individual of a certain type (human being or mosquito).

However, a parameter or state variable that measures the number of individuals of a certain type has a dimension-like quality associated with it. The reason for this statement is that it is meaningful to add two variables if they measure numbers of individuals of the same type, but meaningless to add them if they measure numbers of individuals of different type. We introduce the term "quasi-dimension" as a generalization of dimension in which also types of individuals are accounted for.

For the problems posed by our system of equations we use one fundamental dimensional unit, namely time $\underline{T}$, and three fundamental quasi-dimensional units, namely "human beings" $\underline{H}$, "mosquitoes" $\underline{M}$, and "bites" $\underline{B}$. The parameter $\underline{N}_1$, the total number of human beings, is dimensionless. We shall refer to any dimensionless variable or parameter as having dimension one. The quasi-dimension of $\underline{N}_1$ is $\underline{H}$, since $\underline{N}_1$ gives a count of human beings. In similarity we find that the total number of mosquitoes $\underline{N}_2$ has dimension one and quasi-dimension $\underline{M}$. The human recovery rate $\underline{r}_1$ gives the proportion of human beings that recover in unit time. Its dimension and quasi-dimension are both equal to $\underline{T}^{-1}$. The same holds for the mosquito death rate $\underline{r}_2$. The mosquito man-biting rate $\underline{a}_2$ gives the number of bites per mosquito in unit time. Its dimension is $\underline{T}^{-1}$, while its quasi-dimension id $\underline{B}\ \underline{M}^{-1}\underline{T}^{-1}$. The human infectivity $\underline{b}_1$ counts the number of infected mosquitoes per bite by susceptible mosquitoes on infected human beings. Its dimension is one and its quasi-dimension $\underline{M}\ \underline{B}^{-1}$. The mosquito infectivity $\underline{b}_2$ gives the number of infected human beings per bite by infected mosquitoes on susceptible human beings. Its dimension is one and its quasi-dimension $\underline{H}\ \underline{B}^{-1}$. The two infection probabilities $\underline{p}_1$ and $\underline{p}_2$ have both dimension and quasi-dimension one. The time variable $\underline{t}$ has both dimension and quasi-dimension $\underline{T}$.

The dimension of a product of two or more parameters or state variables is the product of the dimensions of the factors. The corresponding rule holds for evaluation of the quasi-dimension of a product. It follows by applying this rule to the expressions (3.1.3), (3.1.4) for the infection rates $\underline{h}_1(\underline{t})$ and $\underline{h}_2(\underline{t})$ that both of them have quasi-dimension $\underline{T}^{-1}$. A summary of the results of our quasi-dimensional analysis is given in Table 3.1.

The quasi-dimension of any quantity contains more information than its dimension (unless they are equal). The dimension of a quantity is found from its quasi-dimension by replacing all quasi-dimensional fundamental units by one.

The first step in a full quasi-dimensional analysis is to replace each variable $\underline{v}$ by a variable that is free from quasi-dimension. This is done by the identification for each variable $\underline{v}$ of some parameter combination $\underline{p}$ that has the same quasi-dimension as $\underline{v}$. One then introduces $\underline{v}/\underline{p}$ as a new variable, which clearly is free of quasi-dimension.

Table 3.1 Quasi-dimensions of parameters and state variables for the Ross malaria model.

Parameter or state variable	Quasi-dimension
N_1	H
N_2	M
r_1, r_2	T^{-1}
a_2	$B\ M^{-1} T^{-1}$
b_1	$M\ B^{-1}$
b_2	$H\ B^{-1}$
p_1, p_2	1
t	T
h_1, h_2	T^{-1}
T_1, T_2	1

From the given system of equations for the variables $\underline{v}$ one then derives a new system of equations for the new variables $\underline{v}/\underline{p}$. By inspection of this new system of equations one can often easily identify the new parameters that determine the solution as functions of the original ones.

There are usually a number of ways in which such an analysis can be carried out. For example, one can often find several different parameter combinations that all have the same quasi-dimension as a given variable $\underline{v}$. The choice among available alternatives is often guided by subject-matter insight and by trial-and-error. A useful guideline is to search for a parameter combination $\underline{p}$ that in some sense is a typical value ("an intrinsic reference") for the variable $\underline{v}$.

We shall in general not require a full quasi-dimensional analysis of our equations; the time $\underline{t}$ will not be scaled (except in Section 5.3). Our main goal is to reduce the number of parameters that are necessary for a study of the steady-state solutions, i.e. the solutions that are found by setting the right-hand sides of the equations equal to zero. We shall, however, briefly illustrate below how the replacement of the time $\underline{t}$ by a variable free of quasi-dimension can be used for an analysis of the time-dependence of the solution.

We note that both state variables $\underline{p}_1$, $\underline{p}_2$ in the system of equations (3.1.5), (3.1.6) are free of quasi-dimension. No replacement of variables is required for this model; all we have to do is to identify the new parameters. Note that any factor that is common to both the terms in the right-hand side of either of the equations has no influence on the steady-state solutions; it will only serve to influence the time-dependence of the solution. We factor out $\underline{r}_1$ of the right-hand side of equation (3.1.5) and $\underline{r}_2$ of the right-hand side of equation (3.1.6). In the remainder of the right-hand sides we can identify two parameter combinations that are termed transmission factors and are denoted by $\underline{T}_1$ and $\underline{T}_2$. They are defined as functions of the basic parameters by

$$T_1 = \frac{a_2 b_2 N_2}{r_1 N_1} \tag{3.1.8}$$

and

$$T_2 = \frac{a_2 b_1}{r_2} . \tag{3.1.9}$$

The differential equations (3.1.5) and (3.1.6) can be rewritten in the form

$$p_1' = r_1(T_1 p_2(1-p_1) - p_1), \tag{3.1.10}$$

$$p_2' = r_2(T_2 p_1(1-p_2) - p_2). \tag{3.1.11}$$

Clearly, the critical points of this system of equations are determined by only two parameters, namely the transmission factors $\underline{T}_1$ and $\underline{T}_2$. We shall proceed with an analysis of these equations in the next subsection.

We end this subsection by noting that the time variable $\underline{t}$ can also be replaced by a variable free of quasi-dimension. One way to do this is to introduce τ by putting

$$\tau = r_1 t. \tag{3.1.12}$$

It is readily verified that τ is free of quasi-dimension. The derivative of $\underline{p}_i$ with respect to τ is equal to

$$\frac{dp_i}{d\tau} = \frac{1}{r_1}\frac{dp_i}{dt}, \quad i = 1,2. \tag{3.1.13}$$

From equations (3.1.10), (3.1.11) we then get

$$\frac{dp_1}{d\tau} = T_1 p_2(1-p_1) - p_1, \tag{3.1.14}$$

and

$$\varepsilon \frac{dp_2}{d\tau} = T_2 p_1(1-p_2) - p_2, \tag{3.1.15}$$

where

$$\varepsilon = r_1/r_2. \tag{3.1.16}$$

In these equations, all variables and all parameters are free of quasi-dimension.

The parameter ε is positive but small. The equations (3.1.14), (3.1.15) are therefore in a form where singular perturbation theory can be applied for a study of the time-dependence of the solution. We shall not pursue the analysis in this direction, but concentrate on the qualitative features of the solutions, which can be derived from the system of equations (3.1.10), (3.1.11).

3.1.3 The Critical Points and Their Local Stability

It does not appear possible to solve the system of differential equations (3.1.10), (3.1.11) explicitly. In spite of this, we shall be able to derive some important properties of the solution by using results from the qualitative theory of differential equations. Elementary introductions to this theory are given by Brauer and Nohel (1969) and by Braun (1975). The critical points of the system of differential equations play an important role in our analysis. They correspond to stationary or steady-state solutions. Only those critical points for which both $\underline{p}_1$ and $\underline{p}_2$ lie in the closed unit interval [0,1] are of epidemiological interest. It is shown in Subsection 3.1.4 that our assumption that the initial point belongs to the unit square $\underline{S}$ defined in (3.1.7) implies that the solution $(\underline{p}_1(\underline{t}),\underline{p}_2(\underline{t}))$ belongs to $\underline{S}$ for each $t \geq 0$. Thus, only critical points that lie in $\underline{S}$ can be reached or approached by solutions to the initial-value problem posed by the system of differential equations (3.1.10), (3.1.11) with an initial point in $\underline{S}$.

In this subsection we investigate how many critical points that lie in $\underline{S}$ and what their coordinates are. We shall find that the answers to both these questions are determined by the values of the two transmission factors $\underline{T}_1$ and $\underline{T}_2$. In addition, we study the local stability of the critical points. An outline of a global analysis is given in Subsection 3.1.4.

In the global analysis of the next subsection we make use of the isoclines, the curves in the $\underline{p}_1$-$\underline{p}_2$-plane where the derivatives of the two state variables $\underline{p}_1$ and $\underline{p}_2$ are equal to zero. The isocline where $\underline{p}_1' = 0$ is found from (3.1.10) to be given by

$$p_2 = F_1(p_1) = \frac{1}{T_1}\,\frac{p_1}{1-p_1} \,. \qquad (3.1.17)$$

The isocline where $\underline{p}_2' = 0$ is found from (3.1.11) to be given by

$$p_2 = F_2(p_1) = \frac{T_2 p_1}{1+T_2 p_1} \,. \qquad (3.1.18)$$

Critical points are found as intersections of the two isoclines. The origin is clearly a critical point for all positive values of $\underline{T}_1$ and $\underline{T}_2$. An additional critical point $\bar{\underline{p}}^1$ is found to exist when $\underline{T}_1\underline{T}_2 > 1$. The coordinates of the additional critical point are denoted by $\underline{P}_1$ and $\underline{P}_2$. They are determined by $\underline{T}_1$ and $\underline{T}_2$ as follows:

$$P_1(T_1,T_2) = \frac{T_1T_2-1}{T_1T_2+T_2} \,, \quad T_1T_2 \geq 1, \qquad (3.1.19)$$

and

$$P_2(T_1,T_2) = \frac{T_1T_2-1}{T_1T_2+T_1} \,, \quad T_1T_2 \geq 1. \qquad (3.1.20)$$

We proceed to an analysis of the local stability of the critical points. A short summary of the theoretical background for this analysis is given in Appendix IV. The first step consists in an evaluation of the Jacobian matrix of the transformation $\underline{f}(\underline{p})$ defined by the right-hand side of (3.1.10), (3.1.11). Here, we use $\underline{p}(\underline{t})$ to denote the two-dimensional vector-valued function of $\underline{t}$ whose components are $\underline{p}_1(\underline{t})$ and $\underline{p}_2(\underline{t})$. We find

$$\frac{\partial f}{\partial p}(p)=\begin{pmatrix} -r_1(T_1p_2+1) & r_1T_1(1-p_1) \\ r_2T_2(1-p_2) & -r_2(T_2p_1+1) \end{pmatrix}. \tag{3.1.21}$$

The preceding analysis has shown that the origin with $\underline{p}_1 = \underline{p}_2 = 0$ is a critical point for all parameter values. The Jacobian matrix evaluated at the origin is equal to

$$\frac{\partial f}{\partial p}(0) = \begin{pmatrix} -r_1 & r_1T_1 \\ r_2T_2 & -r_2 \end{pmatrix}. \tag{3.1.22}$$

The stability of the origin is determined by the eigenvalues of this matrix. They satisfy the equation

$$\lambda^2 + (r_1+r_2)\lambda + r_1r_2(1 - T_1T_2) = 0. \tag{3.1.23}$$

An application of the Routh-Hurwitz criterion (AIV.17) shows that both roots of this equation have negative real parts if and only if $\underline{T}_1\underline{T}_2 < 1$. We conclude that the origin is asymptotically stable if $\underline{T}_1\underline{T}_2 < 1$ and unstable if $\underline{T}_1\underline{T}_1 > 1$.

The Jacobian matrix (3.1.21), evaluated at the additional critical point $\bar{\underline{p}}^1$ with the aid of the expressions (3.1.19), (3.1.20) for the components of $\bar{\underline{p}}^1$, is found to be equal to

$$\frac{\partial f}{\partial p}(\bar{p}^1) = \begin{pmatrix} -\dfrac{r_1(T_1+1)T_2}{T_2+1} & \dfrac{r_1T_1(T_2+1)}{(T_1+1)T_2} \\ \dfrac{r_2(T_1+1)T_2}{T_1(T_2+1)} & -\dfrac{r_2T_1(T_2+1)}{T_1+1} \end{pmatrix}. \tag{3.1.24}$$

The eigenvalues of this matrix satisfy the equation

$$\lambda^2 + \left[\frac{r_1(T_1+1)T_2}{T_2+1} + \frac{r_2T_1(T_2+1)}{T_1+1}\right]\lambda + (T_1T_2-1)r_1r_2 = 0. \quad (3.1.25)$$

The criterion in (AIV.17) shows that the critical point $\bar{p}^1$ is asymptotically stable for $T_1T_2 > 1$.

The global analysis of the next subsection will show that every solution of our system of differential equations approaches a critical point as t approaches infinity. The relation $T_1T_2 = 1$ determines a threshold function in parameter space, i.e. in the positive quadrant where the transmission factors T_1 and T_2 take their values. Epidemiological interpretations of the threshold function are given in the next subsection.

3.1.4 Global Analysis

We proceed to give an outline of a global analysis of the system of differential equations (3.1.10), (3.1.11). Our treatment is heuristic. We illustrate the solution properties by the aid of so-called phase portraits. All of the results quoted can be proved rigorously. The phase portraits are useful for the formulation of the rigorous treatment. A brief description of theoretical concepts follows.

The time t does not appear explicitly on the right-hand side of our system of equations. The system is therefore said to be <u>autonomous</u>. Any solution of an autonomous system of two differential equations can be represented by a curve, called <u>orbit</u>, in the <u>phase plane</u>, which in our case is the p_1-p_2-plane. The slope of the orbit at any point in the phase plane is a vector whose components are given by the right-hand sides of the equations. The phase plane with arrows indicating these vectors is called a <u>phase portrait</u> of the system. In drawing the phase portrait it is helpful to identify the <u>isoclines</u> and their points of intersection, the <u>critical points</u>. A subset of the phase plane is called <u>positively invariant</u> if it contains the entire orbit for all $t \geq 0$ through each of its points. In studying a system of differential equations, it suffices to restrict attention to a positively invariant subset that contains all biologically meaningful initial values.

In order to show that a set $\underline{V}$ is positively invariant, we consider the initial value problem $\underline{p}' = \underline{f}(\underline{p})$ with $\underline{p}(0) = \underline{v}$, where $\underline{v}$ is a point on the boundary of $\underline{V}$. The point $\underline{v}$ is defined to be an egress point if the tangent $\underline{T}$ to the orbit of the solution of the initial value problem at the point $\underline{v}$ is pointing outward from the set $\underline{V}$ for increasing time. If a unique normal $\underline{N}$ to the boundary of the set $\underline{V}$ pointing into the set $\underline{V}$ can be defined at the point $\underline{v}$, then $\underline{v}$ is an egress point if the inner product $\underline{N} \cdot \underline{T} < 0$. A point $\underline{v}$ where the boundary of the set $\underline{V}$ is not smooth and where two inward-pointing normals $\underline{N}_1$ and $\underline{N}_2$ can be defined is an egress point if at least one of the inner products $\underline{N}_1 \cdot \underline{T}$ and $\underline{N}_2 \cdot \underline{T}$ is strictly negative. The set $\underline{V}$ is positively invariant if no boundary point is an egress point.

The local stability properties of the critical points have been studied in the preceding subsection. If a critical point is asymptotically stable, then we define its <u>domain of attraction</u> as the subset of the phase plane such that any solution of the system with initial value in the subset approaches the critical point as time approaches infinity. If an asymptotically stable critical point belongs to a positively invariant subset containing all initial values, then so does its domain of attraction. An asymptotically stable critical point is said to be <u>globally stable</u> if its domain of attraction coincides with the smallest positively invariant subset containing all initial values.

The phase portraits for the two cases $\underline{T}_1\underline{T}_2 \leq 1$ and $\underline{T}_1\underline{T}_2 > 1$ are shown in Figures 3.1A and 3.1B. The slopes of the arrows in the phase portraits are not, and cannot be, equal to the orbit slopes. Instead, the arrows are drawn to indicate if $\underline{p}_1$ is increasing or decreasing (i.e. if $\underline{p}_1'$ is positive or negative) and if $\underline{p}_2$ is increasing or decreasing (i.e. if $\underline{p}_2'$ is positive or negative). We note first that the arrows are vertical on the isoclines where $\underline{p}_1' = 0$, and horizontal on the isoclines where $\underline{p}_2' = 0$. Furthermore it follows from (3.1.10) and (3.1.17) that $\underline{p}_1'$ is positive (negative) for $\underline{p}_2 > \underline{F}_1(\underline{p}_1)$ $(\underline{p}_2 < \underline{F}_1(\underline{p}_1))$, and from (3.1.11) and (3.1.18) that $\underline{p}_2'$ is positive (negative) for $\underline{p}_2 < \underline{F}_2(\underline{p}_1)$ $(\underline{p}_2 > \underline{F}_2(\underline{p}_1))$. These simple observations are sufficient to draw the arrows in the two

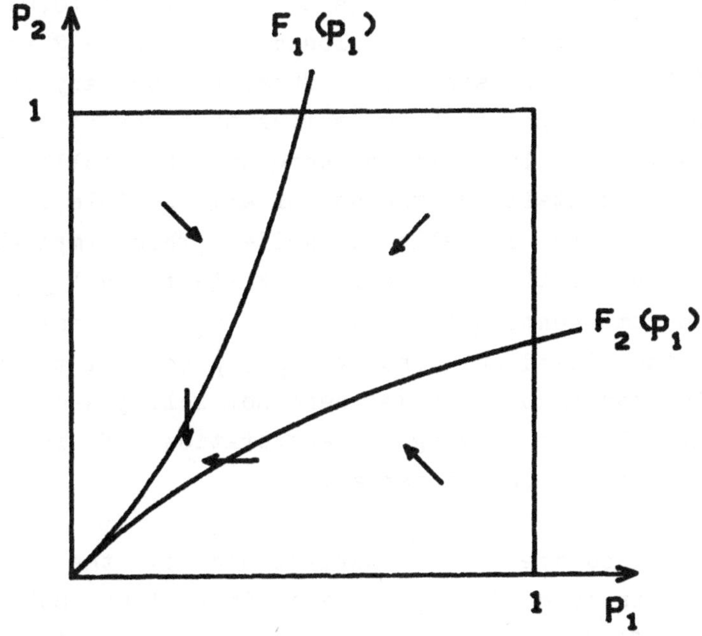

Figure 3.1A. Phase portrait for the Ross malaria model below the threshold.

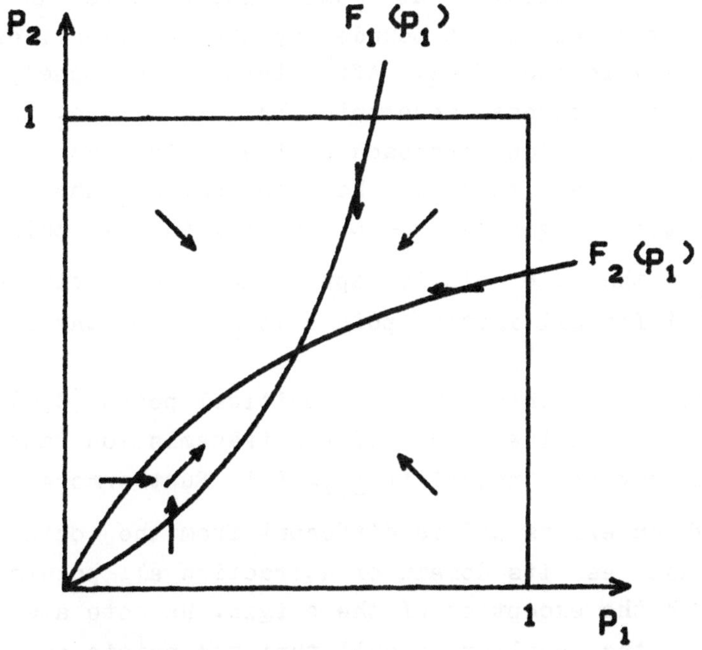

Figure 3.1B. Phase portrait for the Ross malaria model above the threshold.

figures. With the exception of the critical point (0,0), the orbits of all boundary points of the closed unit square $\underline{S}$ point into the interior of $\underline{S}$ for increasing time. Thus, no boundary point of the set $\underline{S}$ is an egress point. This implies that the closed unit square $\underline{S}$ is positively invariant. It is therefore the smallest positively invariant set containing the set of all possible initial values, since the two sets actually coincide. This implies that the solution for any initial value in $\underline{S}$ exists for all $\underline{t} \geq 0$, and that the solution components $\underline{p}_1(\underline{t})$ and $\underline{p}_2(\underline{t})$ both are restricted to take values in the closed unit interval $[0,1]$ for all $\underline{t} \geq 0$. Any other result would mean that the model were not well posed; there is no epidemiologically satisfying interpretation of infection probabilities outside the unit interval.

The closed unit square $\underline{S}$ is partitioned by the isoclines into three subsets in case $\underline{T}_1\underline{T}_2 \leq 1$ and into four subsets in case $\underline{T}_1\underline{T}_2 > 1$. We find that each solution component $\underline{p}_1(\underline{t})$ and $\underline{p}_2(\underline{t})$ is monotonically increasing or monotonically decreasing in each of these subsets. Consider now the case $\underline{T}_1\underline{T}_2 \leq 1$ with an initial point above the isocline $\underline{p}_2 = \underline{F}_1(\underline{p}_1)$. The orbit will then cross this isocline vertically at some finite time and enter the positively invariant subset bounded by the two isoclines and some of the boundary points of $\underline{S}$. After this has happened, both $\underline{p}_1(\underline{t})$ and $\underline{p}_2(\underline{t})$ will decrease monotonically toward zero. The conclusion is that the solution approaches the origin as $\underline{t} \to \infty$. Similar arguments can be used to draw the conclusion that the solution approaches the origin as $\underline{t} \to \infty$ for all initial points in $\underline{S}$ when $\underline{T}_1\underline{T}_2 \leq 1$, and that the solution approaches the critical point $\bar{\underline{p}}^1$ when $\underline{T}_1\underline{T}_2 > 1$ for all initial points in $\underline{S}$ except the origin.

We conclude from this that the critical point (0,0) is globally stable for any positive values of the transmission factors $\underline{T}_1$ and $\underline{T}_2$ that satisfy the inequality $\underline{T}_1\underline{T}_2 \leq 1$. Furthermore, the critical point $\bar{\underline{p}}^1$, which exists and is different from the origin only for $\underline{T}_1\underline{T}_2 > 1$, has as its domain of attraction all points of the unit square $\underline{S}$ with the exception of the origin. We note also that this agrees with the earlier result that the origin is unstable when $\underline{T}_1\underline{T}_2 > 1$.

The set of all solutions is seen to change its behaviour in a qualitative way as we pass the threshold function in the space where the transmission factors take their values. This phenomenon is called <u>bifurcation</u>.

These mathematical results can be given the following epidemiological interpretations: Any amount of infection introduced into a community with $\underline{T}_1\underline{T}_2 \leq 1$ will ultimately die out of its own. The introduction of even a small amount of infection into a community with $\underline{T}_1\underline{T}_2 > 1$ causes the infection to take hold and the infection probabilities among human beings and mosquitoes to rise toward the values $\underline{P}_1(\underline{T}_1,\underline{T}_2)$ and $P_2(\underline{T}_1,\underline{T}_2)$, respectively. They are the endemic infection probabilities that are approached after malaria transmission has been going on for a long time under constant biological and environmental influences.

This means that only communities above the threshold, i.e. with $\underline{T}_1\underline{T}_2 > 1$, can support an endemic infection level. This threshold result implies that eradication of malaria infection in a community is achieved if the transmission factors $\underline{T}_1$ and/or $\underline{T}_2$ are permanently reduced so that the product of their new values is at most equal to one. The product $\underline{T}_1\underline{T}_2$, if larger than one, is a measure of the amount of parameter modification that is required in order to achieve eradication. We refer to $\underline{T}_1\underline{T}_2$ as the eradication effort.

The product $\underline{T}_1\underline{T}_2$ corresponds to what Macdonald (1952) has called the basic reproduction rate; see the comments at the end of Subsection 3.3.3. This quantity can be defined as the average number of secondary cases caused by a single primary infection in a large population of susceptibles. The condition that the number of susceptibles is large implies that the density dependencies in the model can be ignored. The density dependencies serve to limit and determine the infection level whenever the basic reproduction rate exceeds the value one.

3.1.5 Epidemiological Measures in the Endemic Case

We proceed to consider an endemic malaria situation. Thus, our discussion is confined to the subset of parameter space which is above threshold, or, equivalently, to those communities for which the corresponding transmission factors $\underline{T}_1$ and $\underline{T}_2$ have a product larger than one.

The task of obtaining accurate numerical estimates of parameters and state variables may involve considerable practical difficulties. The model results can, however, be used to illuminate or facilitate the estimation problem. Note first that is not necessary to find estimates of all seven basic parameters; the main epidemiological information is contained in the two transmission factors $\underline{T}_1$ and $\underline{T}_2$. Note furthermore that the expressions (3.1.19) (3.1.20) can be solved for $\underline{T}_1$ and $\underline{T}_2$ in terms of the endemic infection probabilities $\underline{P}_1$ and $\underline{P}_2$. The result is

$$T_1 = P_1/(P_2(1-P_1)), \quad 0 < P_1, P_2 < 1 \qquad (3.1.26)$$

and

$$T_2 = P_2/(P_1(1-P_2)), \quad 0 < P_1, P_2 < 1. \qquad (3.1.27)$$

These relations can clearly be used to derive estimates of the transmission factors $\underline{T}_1$ and $\underline{T}_2$ from estimates of the endemic infection probabilities $\underline{P}_1$ and $\underline{P}_2$. The fact that such useful relations can be derived in the simple Ross model should encourage us to seek similar relations in more realistic, and hence more complex, models. This comment applies to all the results derived in this section.

From expression (3.1.19) one finds that

$$T_1 = (1 + P_1T_2)/((1-P_1)T_2), \quad 0 < P_1 < 1, \quad T_2 > 0. \qquad (3.1.28)$$

The right-hand side of this relation is a decreasing function of $\underline{T}_2$ with the asymptotic behaviour $\underline{P}_1/(1-\underline{P}_1)$ as $\underline{T}_2 \to \infty$. We conclude that the following inequalities hold between the transmission factor $\underline{T}_1$ and the endemic infection probability $\underline{P}_1$:

$$T_1 > P_1/(1-P_1), \quad 0 < P_1 < 1, \tag{3.1.29}$$

$$P_1 < T_1/(1+T_1), \quad T_1 > 0. \tag{3.1.30}$$

The infection rate of human beings in steady state is denoted by $\underline{H}_1$. We find from (3.1.3), (3.1.8) and (3.1.20) that it is determined by the three parameters $\underline{u}_1, \underline{r}_1, \underline{T}_2$, where

$$u_1 = \frac{a_2 b_2 N_2}{N_1} = r_1 T_1, \tag{3.1.31}$$

and that it obeys the expression

$$H_1(u_1, r_1, T_2) = r_1 T_1 P_2 = (u_1 T_2 - r_1)/(T_2+1), \quad u_1 T_2 \geq r_1. \tag{3.1.32}$$

This function measures the risk of infection for human beings in steady state and may therefore also be referred to as the steady state public health factor.

The prevalence, incidence and recovery probability in steady state (i.e. with constant infection rate) are given in (2.1.29) - (2.1.31). For the human population we get the prevalence

$$Q(s; u_1, r_1, T_2) = \frac{H_1}{H_1 + r_1} \left(1 - e^{-(H_1 + r_1)s}\right), \tag{3.1.33}$$

the incidence

$$I(s, T; u_1, r_1, T_2) = Q(T; u_1, r_1, T_2), \tag{3.1.34}$$

and the recovery probability

$$R(s, T; u_1, r_1, T_2) = \frac{r_1}{H_1} Q(T; u_1, r_1, T_2). \tag{3.1.35}$$

Insertion of the expression (3.1.32) for the steady state infection rate $\underline{H}_1$ into these relations gives the prevalence, the incidence and the recovery probability defined above the threshold $\underline{u}_1 \underline{T}_2 = \underline{r}_1$ in the space where the three parameters $\underline{u}_1$, $\underline{T}_2$, and $\underline{r}_1$ take their values.

From (3.1.19), (3.1.31) and (3.1.32) we find that

$$P_1 = \frac{H_1}{H_1+r_1} . \tag{3.1.36}$$

Thus, (3.1.33) shows that the prevalence $\underline{Q}(\underline{s})$ approaches the steady state infection probability $\underline{P}_1$ as the age $\underline{s} \to \infty$. The steady state infection probability $\underline{P}_1$ can be interpreted as the prevalence among old human hosts. We note also that the steady state public health factor can be found from the incidence or the prevalence through the relation

$$H_1 = \frac{\partial}{\partial T} I(s,0) = Q'(0). \tag{3.1.37}$$

Thus, the steady state public health factor is equal to the slope at $\underline{T} = 0$ of the incidence as a function of $\underline{T}$, and also equal to the slope at $\underline{s} = 0$ of the prevalence as a function of the age $\underline{s}$.

3.1.6 Control Efficiencies

Control of a malaria infection in a community corresponds to permanent modifications of the values of one or more of the basic parameters. Of course, the parameter modifications considered for control are those that lead to an improvement of the health condition in the community. Above we have introduced a number of epidemiological quantities that all can be used as indicators of the health condition in the community. Thus, the effect of any possible control action can be measured in a number of different ways. We shall introduce functions on parameter space that measure the efficiency of control of various epidemiological quantities.

The parameter space will be either the one where the two transmission factors $\underline{T}_1$ and $\underline{T}_2$ take their values, or the one where the three parameters $\underline{u}_1$, $\underline{r}_1$, $\underline{T}_2$ take their values. In judging the effect of a control action, we shall only by concerned with the longterm effect on the steady-state value of the epidemiological quantity in question. The time necessary for the change to take place does not enter into the analysis.

The following five epidemiological quantities will be used as indicators of the health condition in the community:

1. The steady state infection probability $\underline{P}_1(\underline{T}_1,\underline{T}_2)$.
2. The steady state public health factor $\underline{H}_1(\underline{u}_1,\underline{r}_1,\underline{T}_2)$.
3. The steady state prevalence $\underline{Q}(\underline{s};\ \underline{u}_1,\underline{r}_1,\underline{T}_2)$ at age $\underline{s}$.
4. The steady state incidence $\underline{I}(\underline{s},\underline{T};\ \underline{u}_1,\underline{r}_1,\underline{T}_2)$ at age $\underline{s}$ over the time interval $\underline{T}$.
5. The steady state recovery probability $\underline{R}(\underline{s},\underline{T};\ \underline{u}_1,\underline{r}_1,\underline{T}_2)$ at age $\underline{s}$ over the time interval $\underline{T}$.

These five quantities are not independent. We have noted above that the prevalence at age $\underline{s}$, $\underline{Q}(\underline{s})$, approaches the infection probability $\underline{P}_1$ as the age $s \to \infty$. The prevalence $\underline{Q}(\underline{s})$, the incidence $\underline{I}(\underline{s},\underline{T})$, and the recovery probability $\underline{R}(\underline{s},\underline{T})$ are always related through relation (2.2.36). The public health factor $\underline{H}_1$ is the derivative of the incidence $\underline{I}(\underline{s},\underline{T})$ with respect to $\underline{T}$, evaluated at $\underline{T} = 0$. Further relations are derived from the result that the incidence $\underline{I}(\underline{s},\underline{T})$ at age $\underline{s}$ over the time interval $\underline{T}$ is equal to the prevalence $\underline{Q}(\underline{T})$ at age $\underline{T}$.

Heuristically, we expect any control action to lead to a decrease in the first four quantities, but to an increase in the last one, the recovery probability.

The efficiency of control of the steady state infection probability $\underline{P}_1$ through a reduction of the transmission factor $\underline{T}_1$ can be measured e.g. by the partial derivative $\dfrac{\partial P_1}{\partial T_1}$. However, the control efficiency measure that we shall use in this case is $\underline{T}_1 \dfrac{\partial P_1}{\partial T_1}$. The reason for preferring this form lies in the way in which the transmission factors have been defined. Notice from (3.1.8) that $\underline{T}_1$ is proportional to $\underline{N}_2$ (the number of mosquitoes). Assume that $\underline{N}_2$ is reduced to $\underline{N}_2' = \underline{N}_2 - \Delta\underline{N}_2$, and denote the corresponding values of the transmission factor by $\underline{T}_1$ and $\underline{T}_1' = \underline{T}_1 - \Delta\underline{T}_1$, respectively. It follows then that

$$\frac{\Delta T_1}{T_1} = \frac{\Delta N_2}{N_2} , \qquad (3.1.38)$$

i.e. the proportion by which T_1 is reduced is the same as the proportion by which N_2 is reduced. A correspondingly simple relation does not hold between the magnitudes of the reductions (ΔT_1 and ΔN_2). The control efficiency function defined in this way is approximately equal to the percent reduction in P_1 that results from a one percent reduction in T_1. We note that our control efficiency function is quasi-dimensionless and nonnegative. We shall define additional control efficiency functions in such a way that these two properties are satisfied.

The efficiency of control of an epidemiological quantity A through modification of a parameter B is measured by a control efficiency function, which we denote by C_{AB} . In all those cases where A is a probability or a proportion, and hence quasi-dimensionless, we define

$$C_{AB} = \pm B \frac{\partial A}{\partial B} , \qquad (3.1.39)$$

where the sign is chosen to make C_{AB} nonnegative. In other cases (which in the present model appear only when $A = H_1$), we define

$$C_{AB} = \pm \frac{B}{A} \frac{\partial A}{\partial B} , \qquad (3.1.40)$$

where again the sign is chosen to make C_{AB} nonnegative.

Consideration of the five epidemiological quantities listed above for judging the effect of a control action leads to fourteen control efficiency functions; we can define two control efficiency functions for the steady state infection probability P_1 and three for each of the remaining four indicators. We deal here only with the two functions that measure the efficiencies of control of P_1 through reductions of T_1 and T_2, respectively. For brevity they are denoted by C_1 and C_2. The remaining twelve control efficiency functions are treated in Appendix III.

Explicit expressions for the control efficiency functions C_1 and C_2 are found after differentiation of (3.1.19). The results are

$$C_1(T_1,T_2) = T_1 \frac{\partial P_1}{\partial T_1} = \frac{T_1(T_2+1)}{(T_1+1)^2 T_2} \qquad (3.1.41)$$

and

$$C_2(T_1,T_2) = T_2 \frac{\partial P_1}{\partial T_2} = \frac{1}{(T_1+1)T_2} \, . \qquad (3.1.42)$$

Certain bounds for the control efficiencies $\underline{C}_1$ and $\underline{C}_2$ can be established. Viewed as a function of $\underline{T}_1$ for constant $\underline{T}_2$, $\underline{C}_1$ has a maximum value of $(1 + 1/\underline{T}_2)/4$, achieved at $\underline{T}_1 = 1$. With constant $\underline{T}_1$, $\underline{C}_1$ is a monotonically decreasing function of $\underline{T}_2$. For large values of $\underline{T}_2$, $\underline{C}_1$ approaches the value $\underline{T}_1/(\underline{T}_1+1)^2$. Thus we conclude that

$$T_1/(T_1+1)^2 < C_1 < (1 + 1/T_2)/4. \qquad (3.1.43)$$

The control efficiency function $\underline{C}_2$ is a monotonically decreasing function of both $\underline{T}_1$ and of $\underline{T}_2$. As $\underline{T}_1$ decreases toward zero, $\underline{C}_2$ approaches the value $1/\underline{T}_2$. Thus we have established an upper bound for $\underline{C}_2$:

$$0 < C_2 < 1/T_2. \qquad (3.1.44)$$

The carrying out of a control action is equivalent with a permanent lowering of at least one of the values of $\underline{T}_1$ and $\underline{T}_2$. Our results allow us to conclude that any control action taken leads, with one exception, to increased values of the control efficiency functions $\underline{C}_1$ and $\underline{C}_2$. The exception is that $\underline{C}_1$ will actually be decreased if $\underline{T}_1 < 1$ and if the amount by which $\underline{T}_1$ is lowered is sufficiently much larger than the amount by which $\underline{T}_2$ is lowered.

The ratio between the control efficiency functions $\underline{C}_1$ and $\underline{C}_2$ can be written

$$\frac{C_1}{C_2} = \frac{1}{1-P_2} > 1. \qquad (3.1.45)$$

It follows from this expression that the control efficiency $\underline{C}_1$ is larger than the control efficiency $\underline{C}_2$ whenever the steady state mosquito prevalence $\underline{P}_2$ is larger than 0, i.e. whenever $\underline{T}_1\underline{T}_2 > 1$.

Thus, in any community above threshold, a given proportionate reduction of $\underline{T}_1$ leads to a larger reduction of $\underline{P}_1$ than an equally large proportionate reduction of $\underline{T}_2$. The advantage of $\underline{T}_1$ over $\underline{T}_2$ increases as $\underline{P}_2$ becomes larger. Near the threshold $\underline{P}_2$ is small and the advantage is small.

The generality inherent in the result that $\underline{C}_1$ is larger than $\underline{C}_2$ is underscored. By directing our study toward qualitative questions, we are able to derive results that hold for all communities, regardless of the particular numerical parameter values that characterize each community.

By using (3.1.19) we find that the control efficiency function $\underline{C}_1$ can be written

$$C_1 = T_1(1 - P_1)/(T_1 + 1) < 1 - P_1. \qquad (3.1.46)$$

Thus, the control efficiency $\underline{C}_1$ is small if the steady state infection probability $\underline{P}_1$ is large.

The ideas used in this discussion and in Appendix III can readily be extended to deal with sensitivities with regard to the basic parameters instead of the transmission factors. We omit details.

3.2. A Malaria Model with Superinfection

About 40 years passed after Ross published his model before new ideas on malaria modelling appeared in the literature, Macdonald (1950). A series of papers by the same author in the 1950's culminated in a textbook, Macdonald (1957). A selection of the papers is contained in a book, published posthumously, Macdonald (1973). Macdonald used his model to establish a theory of malaria eradication. As in the Ross model, the conditions for eradication are found from a threshold phenomenon associated with bifurcation in a nonlinear system of differential equations. Macdonald's work on malaria models has had a large influence on the thinking of epidemiologists and public health workers, and it has found important applications in the malaria eradication programme launched by the World Health Organization in the 1960's.

Macdonald modified and extended the Ross model in two ways. Thus, the Macdonald model makes allowance for multiple infections ("superinfections") of human beings and for the latent period ("the extrinsic cycle") in the mosquito. The malaria transmission model treated in this section accounts for the superinfection but not for the extrinsic cycle. Comments on how to deal with the latter phenomenon are given in Section 3.3. The two phenomena lead to quite different mathematical problems. A hybrid model that accounts for both superinfection among human beings and latency among mosquitoes is easily established as a variation of the model of the present section. A comparison of the properties of the model of the present section with those of the model of the previous section will be made. It will allow us to judge the epidemiological influence of the hypothesis of superinfection.

There is one point at which Macdonald's mathematical treatment of superinfection is inconsistent with the verbal description that he gives of his hypotheses. The inconsistency concerns the rate of termination of infections. The inconsistency was first discovered by Dietz, and mentioned by him in 1970 in an unpublished report. The first published document that refers to the inconsistency is the paper by Dietz, Molineaux, and Thomas (1974). It is also described by Bailey (1975), (1982). The paper by Fine, (1975), is devoted to a detailed description and analysis of Macdonald's inconsistency. Our treatment of superinfection is based on Macdonald's intended (verbal) description of his hypotheses, and not on the mathematical model that he developed. Even though Macdonald's inconsistency has been known for some time, it appears that until now have neither a rigorous formulation of the full malaria transmission model intended by Macdonald been published, nor have the full consequences of the hypotheses that Macdonald formulated been drawn.

3.2.1 Model Formulation

The model of the present section uses seven basic parameters, just as the hybrid version of the Ross model treated in the previous section, and the notation is identical. The only differences are that the parameter r_1 has a different interpretation, and that the definition of the infectivity b_2 of mosquitoes is slightly reworded. These two parameters are defined as follows:

r_1 = the rate of termination of an individual infection in a human host,

b_2 = the infectivity of an infected mosquito, defined as the probability that a bite by an infected mosquito on a human being will transfer infection to the human being.

The model takes the explicit form of a collection of N_1+1 Markov chains as follows:

$$\{X_1^{(1)}(t),\ X_1^{(2)}(t), \ldots,\ X_1^{(N_1)}(t),\ X_2(t)\}, \quad t \geq 0.$$

Here, $X_1^{(k)}(t)$ stands for the number of infections in human being number k at time t, while $X_2(t)$ indicates the number of infected mosquitoes at time t. Each Markov chain $X^{(k)}$ has the state space $\{0,1,2,\ldots\}$, while the Markov chain X_2 has the state space $\{0,1,\ldots,N_2\}$.

Infection of a human being has its counterpart in a transition of the corresponding Markov chain from some state n to the state $n+1$. Such transitions take place with a rate that we denote by $h_1(t)$. Termination of an individual infection in a human being corresponds to a transition of the Markov chain associated with this individual from some state $n \geq 1$ to the state $n-1$. Such a transition takes place with the rate r_1 per infection. This implies that the rate with which such a transition of the Markov chain takes place is nr_1. A transition from n to $n+1$ of the Markov chain X_2 corresponds to infection of one mosquito. The rate per uninfected mosquito at which such a transition takes place is denoted by $h_2(t)$. The death of an infected mosquito corresponds to a transition of the Markov chain X_2 from n to $n-1$. Such transitions take place with the rate r_2 per infected mosquito.

Our treatment of superinfection is based on the following quote from Macdonald (1950): "The existence of infection is no barrier to superinfection, so that two or more broods of organisms may flourish side by side, the duration of infection due to one being unaltered by others". We note that in the mathematical formulation of Macdonald's model, the rate of transition of any Markov chain $\underline{X}_1^{(k)}(\underline{t})$ from $\underline{n}$ to $\underline{n}-1$ was put equal to $\underline{r}_1$, not $\underline{n}\underline{r}_1$, a clear inconsistency.

The superinfection process of Section 2.2 is the host model that applies for all of the human host Markov chains $\underline{X}_1^{(k)}$ in the present transmission model, while the second model of Section 2.6 applies for the mosquito Markov chain $\underline{X}_2$.

We assume that the initial distributions of all human host Markov chains, i.e. the distributions of $\underline{X}_1^{(k)}(0)$, are identical for all hosts $\underline{k}$ and are given by Poisson distribution with common parameter $\underline{x}_1(0)$. It follows from Section 2.2 that each $\underline{X}_1^{(k)}(\underline{t})$ has a Poisson distribution for each $\underline{t} > 0$ with a time-dependent paramemeter $\underline{x}_1(t)$. Furthermore, (2.2.32) shows that the expected number of infections per human being $\underline{x}_1$ satisfies the differential equation

$$x_1' = h_1(t) - r_1 x_1. \tag{3.2.1}$$

The infection probability among the human hosts is denoted by $\underline{p}_1(\underline{t})$. Since $\underline{X}_1^{(k)}(\underline{t})$ has a Poisson distribution with parameter $\underline{x}_1(\underline{t})$ we get

$$p_1(t) = P\{X_1^{(k)}(t) > 0\} = 1 - e^{-x_1(t)}. \tag{3.2.2}$$

For the mosquito phase we assume that $\underline{X}_2(0)$ has a binomial distribution with parameters $\underline{N}_2$ and $\underline{p}_{20}$. This implies that the infection probabilities are the same for all mosquitoes. Denoting each of these by $\underline{p}_2(\underline{t})$ we find from (2.6.5) that $\underline{p}_2$ satifies the differential equation

$$p_2' = h_2(t)(1 - p_2) - r_2 p_2. \tag{3.2.3}$$

By using the same arguments as in Section 3.1, we find that the infection rates $\underline{h}_1(\underline{t})$ and $\underline{h}_2(\underline{t})$ can be expressed in terms of the state variables $\underline{x}_1$, $\underline{p}_2$ and the parameters as follows:

$$h_1(t) = \frac{a_2 b_2 N_2}{N_1} p_2(t), \tag{3.2.4}$$

$$h_2(t) = a_2 b_1 (1 - e^{-x_1}). \tag{3.2.5}$$

Thus we are led to the following system of differential equations for the state variables $\underline{x}_1$ and $\underline{p}_2$:

$$x_1' = \frac{a_2 b_2 N_2}{N_1} p_2 - r_1 x_1, \tag{3.2.6}$$

$$p_2' = a_2 b_1 (1 - e^{-x_1})(1 - p_2) - r_2 p_2. \tag{3.2.7}$$

3.2.2 Quasi-Dimensional Analysis

We proceed to analyze the quasi-dimensions of parameters and state variables in these equations. In addition to the fundamental dimensional unit of time $\underline{T}$, and the three quasi-dimensional fundamental units of human beings $\underline{H}$, of mosquitoes $\underline{M}$, and of bites $\underline{B}$, introduced in Section 3.1, we also need the quasi-dimensional fundamental unit of "infections of human beings" $\underline{I}$. Indeed, the infectivity $\underline{b}_2$ of mosquitoes measures the proportion of bites (by infected mosquitoes) that result in infections. Its quasi-dimension is therefore $I\ B^{-1}$. The remaining six basic parameters have the same quasi-dimensions as in the Ross model; see Table 3.2.1. The state variable $\underline{x}_1$ measures the expected number of infections per human host and has quasi-dimension $\underline{I}\ \underline{H}^{-1}$. The human infection rate is found from (3.2.4) to have the quasi-dimension $\underline{I}\ \underline{H}^{-1}\underline{T}^{-1}$. A summary of the results of the quasi-dimensional analysis is given in Table 3.2.1.

We note that the state variable $\underline{x}_1$ has quasi-dimension $\underline{IH}^{-1}$, while the state variable $\underline{p}_2$ is already free of quasi-dimension. To replace $\underline{x}_1$ by a variable free of quasi-dimension we need to find a

Table 3.2.1 Quasi-dimensions of parameters and state variables for the malaria model with superinfection.

Parameter or state variable	Quasi-dimension
N_1	H
N_2	M
r_1, r_2	T^{-1}
a_2	$B\ M^{-1}T^{-1}$
b_1	$M\ B^{-1}$
b_2	$I\ B^{-1}$
p_1, p_2	1
x_1	$I\ H^{-1}$
t	T
h_1	$I\ H^{-1}T^{-1}$
h_2	T^{-1}
T_1	$I\ H^{-1}$
T_2	1

parameter combination whose quasi-dimension is the same as that of $\underline{x}_1$, i.e. equal to $\underline{IH}^{-1}$. There are many parameter combinations that satisfy this requirement, and this means, as remarked before, that several different alternatives are available. If $\underline{r}_1$ is factored out of the right-hand side of equation (3.2.6) and if $\underline{r}_2$ is factored out of the right-hand side of equation (3.2.7), then we find that the following parameter combinations appear as factors in the first term of each of the two equations:

$$T_1 = \frac{a_2 b_2 N_2}{r_1 N_1}, \tag{3.2.8}$$

$$T_2 = \frac{a_2 b_1}{r_2}. \tag{3.2.9}$$

We note that these definitions of transmission factors are formally identical vith the definitions of $\underline{T}_1$ and $\underline{T}_2$ in (3.1.8) and (3.1.9) for the Ross model. We note also from (3.2.6) that $\underline{T}_1$ is a "typical value" for $\underline{x}_1$ in the following sense: If all mosquitoes are infected (i.e. if $\underline{p}_2 = 1$), then $\underline{x}_1$ will approach the value $\underline{T}_1$ as time grows toward infinity. In analogy with the terminology in dimensional analysis we refer to $\underline{T}_1$ as an intrinsic reference for the expected number of infections per human host. Equation (3.2.7) shows that the mosquito infection probability $\underline{p}_2$ approaches the value $\underline{T}_2/(1+\underline{T}_2)$ if the expected number of infections per human being $\underline{x}_1$ is large.

The quasi-dimension of $\underline{T}_1$ is $\underline{I}\ \underline{H}^{-1}$, while that of $\underline{T}_2$ is one. Thus, the transmission factor $\underline{T}_1$ can be used to replace the state variable $\underline{x}_1$ by a new variable $\underline{z}_1$ defined by the expression

$$z_1 = x_1/T_1. \tag{3.2.10}$$

For notational convenience we introduce

$$z_2 = p_2. \tag{3.2.11}$$

It is clear that both $\underline{z}_1$ and $\underline{z}_2$ are free of quasi-dimension. From equation (3.2.6), (3.2.7) we find the following system of equations for the variables $\underline{z}_1$, $\underline{z}_2$:

$$z_1' = r_1\ (z_2 - z_1), \tag{3.2.12}$$

$$z_2' = r_2\ (T_2(1 - e^{-T_1 z_1})(1 - z_2) - z_2). \tag{3.2.13}$$

We note that the exponent of $\underline{e}$ in the right-hand side of equation (3.2.13) is dimensionless, as it must be, but not free of quasi-dimension.

Initial values for the variables $\underline{z}_1$ and $\underline{z}_2$ are specified at $\underline{t} = 0$ by $\underline{z}_{10} = \underline{x}_{10}/\underline{T}_1$ and $\underline{z}_{20} = \underline{p}_{20}$. The initial point $(\underline{z}_{10}, \underline{z}_{20})$ belongs to the rectangle

$$R = \{(z_1, z_2) : 0 \leq z_1 \leq \hat{z}_1, \quad 0 \leq z_2 \leq 1\}, \tag{3.2.14}$$

where $\hat{\underline{z}}_1 = \max(1, \underline{z}_{10})$. It is shown below that the orbit of every solution with initial value in $\underline{R}$ belongs to $\underline{R}$ for all $\underline{t} \geq 0$.

3.2.3 Model Analysis

We begin the model analysis with a study of the isoclines of the system (3.2.12), (3.2.13). The isocline where $\underline{z}_1' = 0$ is given by a straight line through the origin with slope one:

$$z_2 = F_1(z_1) = z_1. \tag{3.2.15}$$

The isocline where $\underline{z}_2' = 0$ is the curve

$$z_2 = F_2(z_1) = \frac{T_2(1 - e^{-T_1 z_1})}{1 + T_2(1 - e^{-T_1 z_1})}, \tag{3.2.16}$$

which also passes through the origin. Critical points are found as intersections of the isoclines. We find that the $\underline{z}_1$-component of any critical point satisfies the equation

$$\bar{z}_1 = T_2(1 - e^{-T_1 \bar{z}_1})(1 - \bar{z}_1). \tag{3.2.17}$$

Note first that $\bar{\underline{z}}_1 = 0$ is a solution of this equation. Thus the origin is a critical point for all positive values of the two transmission factors. This is to be expected, since, as remarked above, both isoclines pass through the origin. For any other critical point we have $\bar{\underline{z}}_1 \neq 0$. Hence we can write

$$T_2 = \frac{\bar{z}_1}{(1-\bar{z}_1)(1 - e^{-T_1 \bar{z}_1})}. \tag{3.2.18}$$

Our immediate goal is to solve this equation for $\bar{\underline{z}}_1$ as a function of $\underline{T}_1$ and $\underline{T}_2$. As a preparation for this we define two functions $\underline{f}$ and $\underline{g}$ by the following relations:

$$f(x) = \begin{cases} 1, & x = 0, \\ \dfrac{x}{1-e^{-x}}, & x > 0, \end{cases} \tag{3.2.19}$$

$$g(T_1,z) = \frac{1}{T_1(1-z)} f(T_1z), \quad T_1 > 0, \quad 0 \leq z < 1. \qquad (3.2.20)$$

The function f has already been introduced and studied in Appendix I. The condition (3.2.18) that the z_1-coordinate of any critical point different from the origin must satisfy can now be written in the form

$$T_2 = g(T_1,\bar{z}_1). \qquad (3.2.21)$$

We proceed to study the conditions on the transmission factors T_1 and T_2 under which this equations has a solution. We note first from the definition (3.2.19) of the function f that its derivative f' can be expressed as follows:

$$f'(x) = \frac{f(x)(1 + x - f(x))}{x}, \quad x > 0. \qquad (3.2.22)$$

It follows therefore that we can write

$$f(x) - xf'(x) = f(x)(f(x) - x), \quad x > 0. \qquad (3.2.23)$$

By using this expression we find after differentiation that the partial derivatives of g can be expressed as follows:

$$\frac{\partial g}{\partial T_1} = - \frac{f(T_1z)(f(T_1z) - T_1z)}{T_1^2(1-z)} < 0, \quad T_1 > 0, \ 0 < z < 1, \qquad (3.2.24)$$

$$\frac{\partial g}{\partial z} = \frac{f(T_1z)(1-(1-z)(f(T_1z)-T_1z))}{T_1z(1-z)^2} > 0, \quad T_1 > 0, \ 0 < z < 1. \qquad (3.2.25)$$

The inequalities here follow from the bounds for $f(x)$ in (AI.14). The arguments T_1 and z of g are omitted for brevity.

For each fixed value of $T_1 > 0$ we consider g as a function of z. We note that

$$g(T_1,0) = 1/T_1 \qquad (3.2.26)$$

and that

$$g(T_1,z) \to \infty \quad \text{as} \quad z \to 1. \qquad (3.2.27)$$

Relation (3.2.25) shows that $g(\underline{T}_1,.)$ is a monotonically increasing function of $\underline{z}$ in the open unit interval (0,1). Hence the range of values taken on by $g(\underline{T}_1,\underline{z})$ as $\underline{z}$ runs through the interval [0,1) is the set of all real numbers larger than or equal to $1/\underline{T}_1$.

Now equation (3.2.21) has a solution exactly when there exists a $\bar{\underline{z}}_1$- value such that the right-hand side of (3.2.21) is equal to $\underline{T}_2$. The condition for this is $\underline{T}_2 > 1/\underline{T}_1$ or $\underline{T}_1\underline{T}_2 > 1$. The solution is unique since $g(\underline{T}_1,.)$ is monotonically increasing. The solution is clearly a function of the two transmisssion factors $\underline{T}_1$ and $\underline{T}_2$; we denote it by $\underline{Z}_1(\underline{T}_1,\underline{T}_2)$. $\underline{Z}_1$ is a function which is defined whenever the product of its two arguments equals or exceeds one. We show below that $\underline{Z}_1(\underline{T}_1,\underline{T}_2)$ is equal to the steady-state infection probability among the mosquitoes.

We need some properties of the function $\underline{Z}_1$ in the sequel. It follows from its definition that it satisfies the identity

$$T_2 = g(T_1,\ Z_1(T_1,T_2)),\quad T_1T_2 \geq 1. \tag{3.2.28}$$

Differentiation of this relation shows that the partial derivatives of the function $\underline{Z}_1$ with respect to $\underline{T}_1$ and $\underline{T}_2$ satisfy the following expressions:

$$\frac{\partial Z_1}{\partial T_1} = -\frac{\dfrac{\partial g}{\partial T_1}(T_1,Z_1)}{\dfrac{\partial g}{\partial z}(T_1,Z_1)} > 0,\quad T_1T_2 > 1, \tag{3.2.29}$$

$$\frac{\partial Z_1}{\partial T_2} = \frac{1}{\dfrac{\partial g}{\partial z}(T_1,Z_1)} > 0,\quad T_1T_2 > 1. \tag{3.2.30}$$

The arguments $\underline{T}_1$ and $\underline{T}_2$ of $\underline{Z}_1$ are deleted in these expressions. Both these partial derivatives are functions defined above the threshold in $\underline{T}_1$ - $\underline{T}_2$-space. The positivity of each of them follows from the inequalities for the partial derivatives of g in (3.2.24), (3.2.25). Thus we have shown that $\underline{Z}_1$ is a continuous function defined on and above the threshold $\underline{T}_1\ \underline{T}_2 = 1$, and that $\underline{Z}_1$ is monotonically increasing in each of its arguments. We note that $\underline{Z}_1$ is identically equal to zero on the threshold.

We proceed to derive some simple bounds for the function $\underline{Z}_1$. By using the definition (3.2.20) of the function g in (3.2.28), we find that $\underline{Z}_1$ satisfies the relation

$$T_1T_2(1 - Z_1(T_1,T_2)) = f(T_1Z_1(T_1,T_2)), \quad T_1T_2 \geq 1. \tag{3.2.31}$$

By applying the inequalities

$$x < f(x) < x+1, \qquad x > 0, \tag{3.2.32}$$

from (AI.14), we find from (3.2.31) that $\underline{Z}_1$ satisfies the following two inequalities above the threshold:

$$\frac{T_1T_2 - 1}{T_1T_2 + T_1} < Z_1(T_1,T_2) < \frac{T_2}{T_2 + 1}, \qquad T_1T_2 > 1. \tag{3.2.33}$$

We proceed to analyze the local stability of the critical points of the system of differential equation (3.2.12), (3.2.13). The system can be written in vector form

$$z'(t) = f(z(t)). \tag{3.2.34}$$

The Jacobian matrix of the function $\underline{f}$ is found to be

$$\frac{\partial f}{\partial z}(z) = \begin{pmatrix} -r_1 & r_1 \\ r_2T_1T_2(1-z_2)e^{-T_1z_1} & -r_2T_2(1-e^{-T_1z_1})-r_2 \end{pmatrix}. \tag{3.2.35}$$

The origin is a critical point for the system of differential equations for all positive parameter values. Evaluating the Jacobian matrix at the origin, we find

$$\frac{\partial f}{\partial z}(0) = \begin{pmatrix} -r_1 & r_1 \\ r_2T_1T_2 & -r_2 \end{pmatrix}. \tag{3.2.36}$$

The eigenvalues of this matrix satisfy the equation

$$\lambda^2 + (r_1 + r_2)\lambda + r_1 r_2(1 - T_1T_2) = 0. \qquad (3.2.37)$$

By using the criterion (AIV.17) we find that the origin is asymptotically stable for $\underline{T}_1\underline{T}_2 < 1$, i.e. below the threshold, and unstable for $\underline{T}_1\underline{T}_2 > 1$.

We use $\underline{\bar{z}}^1$ to denote the critical point that exists on and above the threshold, i.e. for $\underline{T}_1\underline{T}_2 > 1$. Thus $\underline{\bar{z}}^1$ has both its components equal to $\underline{Z}_1$. The Jacobian matrix evaluated at this critical point is found to be

$$\frac{\partial f}{\partial z}(\bar{z}^1) = \begin{pmatrix} -r_1 & r_1 \\ r_2T_1(T_2-(T_2+1)Z_1) & -\dfrac{r_2}{1-Z_1} \end{pmatrix}. \qquad (3.2.38)$$

The eigenvalues of this matrix satisfy the equation

$$\lambda^2 + \left(r_1 + \frac{r_2}{1-Z_1}\right)\lambda + r_1r_2\left(\frac{1}{1-Z_1} - T_1T_2 + T_1(T_2+1)Z_1\right) = 0. \quad (3.2.39)$$

The coefficient of λ in this equation is always positive. By using the lower bound of $\underline{Z}_1$ in (3.2.33) we find that the constant term is bounded below as follows:

$$\frac{1}{1-Z_1} - T_1T_2 + T_1(T_2+1)Z_1 > \frac{T_1T_2-1}{T_1+1}. \qquad (3.2.40)$$

We conclude therefore, using criterion (AIV.17), that the critical point $\underline{\bar{z}}^1$ is asymptotically stable above the threshold.

Phase portraits for the system of differential equations (3.2.12), (3.2.13) are shown in Figures 3.2A and 3.2B. Figure 3.2A applies for the case when the product of $\underline{T}_1$ and $\underline{T}_2$ is less than or equal to one ("below the threshold"), and Figure 3.2B illustrates the case when $\underline{T}_1\underline{T}_2 > 1$ ("above the threshold"). The directions of the arrows in the subsets of the rectangle $\underline{R}$ bounded by the isoclines are easily found in a manner similar to that used for the Ross model of the preceding section. The rectangle $\underline{R}$ of possible initial values

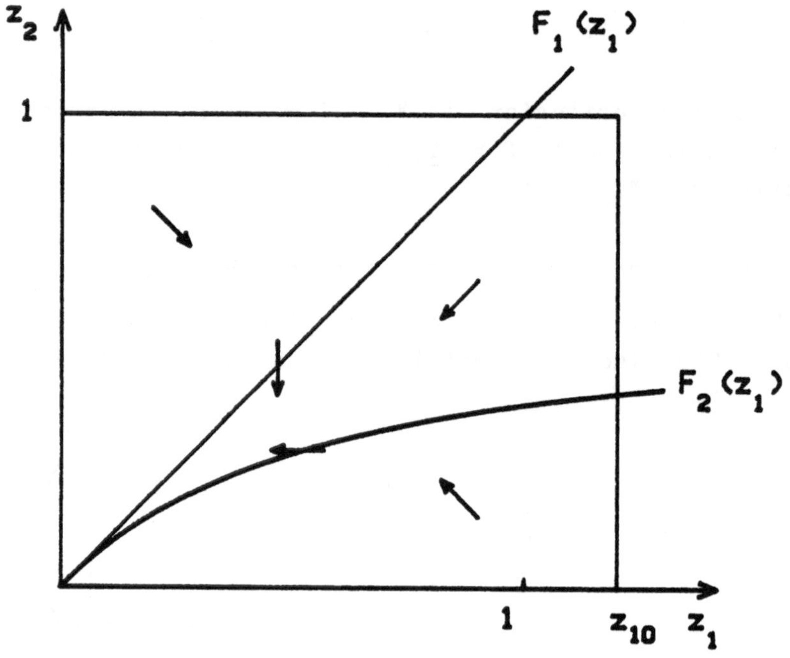

Figure 3.2A. Phase portrait for the malaria superinfection model below the threshold.

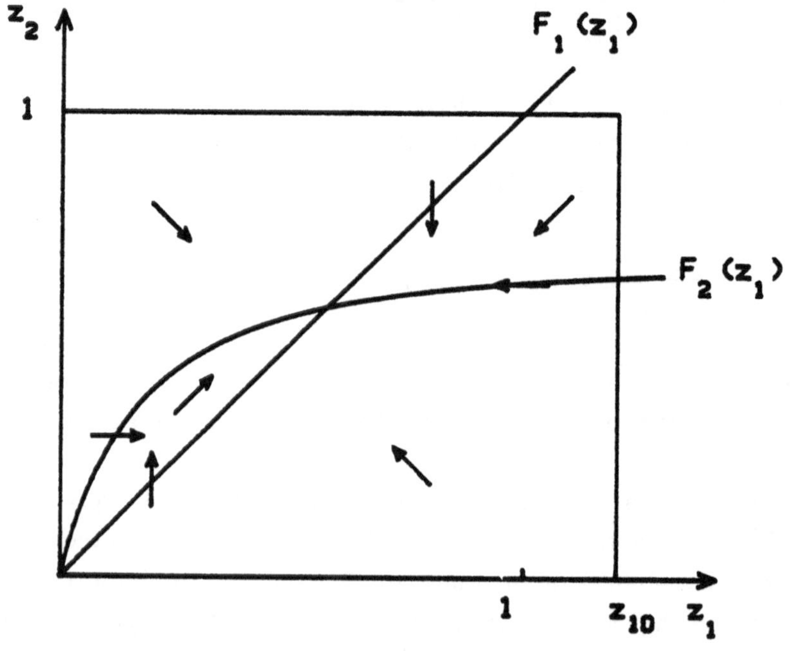

Figure 3.2B. Phase portrait for the malaria superinfection model above the threshold.

defined in (3.2.14) coincides with the unit square whenever the initial value $\underline{z}_{10} \leq 1$. The phase portraits indicate that the rectangle $\underline{R}$ is positively invariant. We conclude therefore that the solution for any initial value in $\underline{R}$ exists for all $\underline{t} \geq 0$, that the solution component $\underline{z}_1(\underline{t})$ takes values in the interval $[0, \max(1,\underline{z}_{10})]$ and that the solution component $\underline{z}_2(\underline{t})$ takes values in the unit interval $[0,1]$. Furthermore, the monotonicity of each solution component within each of the three subsets of $\underline{R}$ shown in Figure 3.2A, and within each of the four subsets of $\underline{R}$ shown in Figure 3.2B implies that every solution approaches a critical point as $\underline{t} \to \infty$. The origin is a globally stable critical point for $\underline{T}_1\underline{T}_2 \leq 1$, but unstable for $\underline{T}_1\underline{T}_2 > 1$. The critical point $(\underline{Z}_1(\underline{T}_1,\underline{T}_2), \underline{Z}_1(\underline{T}_1,\underline{T}_2))$, which exists only for $\underline{T}_1\underline{T}_2 \geq 1$, and is different from the origin only for $\underline{T}_1\underline{T}_2 > 1$, has in the latter case as its domain of attraction all points in the rectangle $\underline{R}$ with the exception of the origin.

3.2.4 The threshold

Epidemiologically, these results mean that there is a threshold function in the parameter space where the two transmission factors take their values. The threshold is determined by the relation $\underline{T}_1\underline{T}_2 = 1$. No endemic infection level can be supported by a community below or on the threshold. Any amount of infection introduced into a community above the threshold will cause an endemic infection level to establish itself. We denote the steady-state expected number of infections per human being by $\underline{X}_1(\underline{T}_1,\underline{T}_2)$ and the steady-state mosquito infection probability by $\underline{P}_2(\underline{T}_1,\underline{T}_2)$. Our results show that they are equal to zero below and on the threshold, and given by

$$X_1(T_1,T_2) = T_1 Z_1(T_1,T_2), \qquad T_1T_2 \geq 1, \tag{3.2.41}$$

$$P_2(T_1,T_2) = Z_1(T_1,T_2), \qquad T_1T_2 \geq 1, \tag{3.2.42}$$

on and above the threshold.

These results form the basis for an eradication theory. An endemic malaria infection in a community will be eradicated if $\underline{T}_1$ and/or

$\underline{T}_2$ are permanently reduced to new values such that the product of the new values is less than or equal to one. The amount of parameter modification required is measured by the product $\underline{T}_1\underline{T}_2$ before eradication measures are applied. The value of $\underline{T}_1\underline{T}_2$ is referred to as the eradication effort. It is equally efficient to reduce $\underline{T}_1$ as to reduce $\underline{T}_2$ in an eradication program.

The fact that the threshold function is the same for the Ross model as for the presently considered model with superinfection should be no surprise. Near the threshold the infection level is small and the mechanism of superinfection will only rarely lead to more than one infection in any human host.

3.2.5 Epidemiological Measures in the Endemic Case

In this subsection we consider a community with endemic malaria. This means that the product of the two transmission factors $\underline{T}_1$ and $\underline{T}_2$ is larger than one. We proceed to evaluate and study expressions for a number of indicators of the steady-state infection level in the community. We begin with the function $\underline{Z}_1$ which by (3.2.42) is equal to the steady-state mosquito infection probability.

The lower bound for $\underline{Z}_1$ in (3.2.33) is shown by (3.1.20) to be equal to the steady-state infection probability among the mosquitoes in the Ross model. Thus, the mechanism of superinfection leads to an increase in the steady-state mosquito infection probability, as can be expected heuristically. Further comparisons between the two models are made in Subsection 3.2.7 below.

Additional information about the properties of the function $\underline{Z}_1$ is gotten from a study of its asymptotic behaviour near the threshold and for large values of $\underline{T}_1$ and $\underline{T}_2$. The argument of $\underline{f}$ in the right-hand side of (3.2.31) is positive everywhere above the threshold. It follows that the right-hand side of (3.2.31) is larger than one above the threshold. We conclude therefore that both $\underline{Z}_1$ and $\underline{T}_1\underline{Z}_1$ are small close to the threshold. By using the asymptotic expression

$$f(x) \sim 1 + \frac{x}{2}, \quad x \to 0 \tag{3.2.43}$$

from (AI.14), we find from (3.2.31) that

$$Z_1(T_1,T_2) \sim \frac{T_1T_2 - 1}{1 + 1/(2T_2)}, \quad T_1 \to 1/T_2 \tag{3.2.44}$$

and that

$$Z_1(T_1,T_2) \sim \frac{T_1T_2 - 1}{1 + T_1/2}, \quad T_2 \to 1/T_1. \tag{3.2.45}$$

We next consider the situation when $\underline{T}_1 \to \infty$ for fixed value of $\underline{T}_2 > 0$. It follows then that $\underline{T}_1\underline{Z}_1 \to \infty$. By using the asymptotic expression

$$f(x) \sim x, \quad x \to \infty \tag{3.2.46}$$

from (AI.14) we find from (3.2.31) that

$$Z_1(T_1,T_2) \sim \frac{T_2}{T_2+1}, \quad T_1 \to \infty . \tag{3.2.47}$$

Finally, we let $\underline{T}_2 \to \infty$ for fixed $\underline{T}_1 > 0$. The right-hand side of (3.2.31) is bounded since (3.2.33) shows that $\underline{Z}_1$ is always smaller than one. It follows therefore from (3.2.31) that $\underline{Z}_1 \to 1$. Thus we conclude that

$$Z_1(T_1,T_2) \sim 1 - \frac{f(T_1)}{T_1T_2}, \quad T_2 \to \infty . \tag{3.2.48}$$

By using expression (3.2.41) for the steady-state expected number of infections per human host we find from (3.2.29) and (3.2.30) that the partial derivatives of $\underline{X}_1$ with respect to both $\underline{T}_1$ and $\underline{T}_2$ are positive:

$$\frac{\partial X_1}{\partial T_1} = Z_1 + T_1 \frac{\partial Z_1}{\partial T_1} > 0, \quad T_1T_2 > 1, \tag{3.2.49}$$

$$\frac{\partial X_1}{\partial T_2} = T_1 \frac{\partial Z_1}{\partial T_2} > 0, \quad T_1T_2 > 1. \tag{3.2.50}$$

Thus we have shown that both the steady-state expected number of infections per human host $\underline{X}_1$ and the steady-state mosquito infection probability $\underline{P}_2$ are increasing functions of both the transmission factors above the threshold, as can be expected heuristically.

The steady-state infection probability among human beings $\underline{P}_1(\underline{T}_1,\underline{T}_2)$ is given by

$$P_1(T_1,T_2) = 1 - e^{-T_1 Z_1(T_1,T_2)} , \qquad T_1T_2 > 1. \qquad (3.2.51)$$

We note that the steady-state infection probabilities $\underline{P}_1$ and $\underline{P}_2$ cannot be expressed as elementary functions of the transmission factors $\underline{T}_1$ and $\underline{T}_2$, since the inverse $\underline{Z}_1$ of the function $\underline{g}(\underline{T}_1,.)$ is not elementary. However, by using relation (3.2.31) and the expressions (3.2.51), (3.2.42) for the steady-state infection probabilites we find that the transmission factors $\underline{T}_1$ and $\underline{T}_2$ can be expressed in terms of the steady-state infection probabilities $\underline{P}_1$ and $\underline{P}_2$ as follows:

$$T_1 = -\frac{\ln(1-P_1)}{P_2} , \qquad 0 < P_1,P_2 < 1, \qquad (3.2.52)$$

$$T_2 = \frac{P_2}{P_1(1-P_2)} , \qquad 0 < P_1,P_2 < 1 . \qquad (3.2.53)$$

These two relations can be used to find estimates of the transmission factors from observed values of the steady-state infection probabilities. We emphasize the conceptual importance of these relations. Similar expressions should be searched for in more realistic models. The model results exhibited here show the importance of knowing the values of the transmission factors.

The two relations (3.2.52) and (3.2.53) hold only when both $\underline{P}_1$ and $\underline{P}_2$ are constrained to lie in the open unit interval (0,1). The restriction that they both must be strictly positive holds above the threshold, which is the case considered in this subsection. On and below the threshold we have $\underline{P}_1 = \underline{P}_2 = 0$, and then there is no way to determine the transmission factors from the infection probabilities. The restrictions that the two steady-state infection probabilities must be strictly less than one are always fulfilled. For $\underline{P}_1$ this follows from (3.2.51), and for $\underline{P}_2$ it follows from (3.2.42) and the implication from (3.2.33) that $\underline{Z}_1$ is always strictly less than one.

We proceed to consider the steady-state infection rate $\underline{H}_1$ of human beings. It will also be referred to as the steady-state public health factor. By using the expression (3.2.4) for the infection rate, the definition (3.2.8) for the transmission factor $\underline{T}_1$, and relation (3.2.42) we find that

$$H_1(u_1,r_1,T_2) = u_1 Z_1(u_1/r_1,\ T_2), \qquad u_1 T_2 \geq r_1. \qquad (3.2.54)$$

Here the parameter $\underline{u}_1$ is defined by

$$u_1 = \frac{a_1 b_2 N_2}{N_1} = r_1 T_1 \qquad (3.2.55)$$

as in (3.1.31). Expression (3.2.54) shows that $\underline{H}_1$ is a function defined on and above the threshold in the space where $\underline{u}_1$, $\underline{r}_1$, $\underline{T}_2$ take their values.

Additional indications of the health condition in the community are given by the steady-state values of the prevalence $\underline{Q}(\underline{s})$, the incidence $\underline{I}(\underline{s},\underline{T})$, and the recovery probability $\underline{R}(\underline{s},\underline{T})$. By using expressions (2.2.33), (2.2.41), (2.2.42), we find the following expressions for these three epidemiological quantities:

$$Q(s;\ u_1,r_1,T_2) = 1 - p_0(s), \qquad (3.2.56)$$

$$I(s,T;\ u_1,r_1,T_2) = 1 - p_0(T), \qquad (3.2.57)$$

$$R(s,T;\ u_1,r_1,T_2) = \frac{p_0(s+T) - p_0(s)p_0(T)}{1 - p_0(s)}. \qquad (3.2.58)$$

Here, $\underline{s}$ denotes the age and $\underline{T}$ denotes the length of the time interval over which incidence and recovery probability are defined. The parameter arguments are indicated for the three epidemiological quantities. The right-hand sides of these three expressions contain a function $\underline{p}_0$ that depends on the same parameter arguments, but this dependence has been omitted from the notation for brevity. The function $\underline{p}_0$ is determined from (2.2.43) and (3.2.54) as follows:

$$p_0(t) = \exp\Big(-(u_1/r_1)Z_1(u_1/r_1,T_2)\Big(1 - e^{-r_1 t}\Big)\Big), \qquad u_1T_2 \geq r_1, \quad t \geq 0. \tag{3.2.59}$$

3.2.6 Control Efficiencies

As was the case for the Ross malaria model, we find that a number of indicators of the steady-state health condition in a community can be used to measure the effect of a control action. We consider here only two of these, namely the expected number of infections per human host $\underline{X}_1$ and the infection probability among human hosts $\underline{P}_1$.

The efficiencies of control of the steady-state expected number of infections per human host $\underline{X}_1$ through reductions of $\underline{T}_1$ and $\underline{T}_2$ are denoted by $\underline{C}_1$ and $\underline{C}_2$, respectively. They are defined above the threshold by

$$C_1 = \frac{T_1}{X_1}\frac{\partial X_1}{\partial T_1} = 1 + \frac{T_1}{Z_1}\frac{\partial Z_1}{\partial T_1}, \quad T_1T_2 > 1, \tag{3.2.60}$$

and

$$C_2 = \frac{T_2}{X_1}\frac{\partial X_1}{\partial T_2} = \frac{T_2}{Z_1}\frac{\partial Z_1}{\partial T_2}, \qquad T_1T_2 > 1. \tag{3.2.61}$$

In order to get explicit expressions for these two control efficiency functions we need to evaluate the partial derivatives of the function $\underline{Z}_1$. These are by (3.2.29) and (3.2.30) determined by the partial derivatives of the function $\underline{g}$ evaluated at $\underline{z} = \underline{Z}_1(\underline{T}_1,\underline{T}_2)$. By applying (3.2.31) in the expressions (3.2.24), (3.2.25) for the partial derivatives for $\underline{g}$, we find that

$$\frac{\partial g}{\partial T_1}(T_1,Z_1) = -\,T_2(T_2 - (T_2+1)Z_1) \tag{3.2.62}$$

and

$$\frac{\partial g}{\partial z}(T_1,Z_1) = \frac{T_2}{Z_1(1-Z_1)}\,(1 - T_1(1-Z_1)(T_2 - (T_2+1)Z_1)). \tag{3.2.63}$$

It follows therefore from (3.2.29), (3.2.30) that the partial derivatives of $\underline{Z}_1$ can be expressed by

$$\frac{\partial Z_1}{\partial T_1} = \frac{Z_1(1-Z_1)(T_2 - (T_2+1)Z_1)}{1 - T_1(1-Z_1)(T_2 - (T_2+1)Z_1)} \tag{3.2.64}$$

and

$$\frac{\partial Z_1}{\partial T_2} = \frac{Z_1(1 - Z_1)}{T_2(1 - T_1(1-Z_1)(T_2 - (T_2+1)Z_1))} \quad . \tag{3.2.65}$$

We therefore conclude that

$$C_1 = \frac{1}{1 - T_1(1-Z_1)(T_2 - (T_2+1)Z_1)} \tag{3.2.66}$$

and that

$$C_2 = (1 - Z_1)C_1. \tag{3.2.67}$$

The ratio between the two control efficiencies is therefore

$$\frac{C_1}{C_2} = \frac{1}{1-P_2} > 1, \tag{3.2.68}$$

formally identical to (3.1.45). We note also from (3.2.66) and (3.2.67) that

$$C_1 = 1 + T_1(T_2-(T_2+1)Z_1)C_2 < 1 + C_2, \tag{3.2.69}$$

where the inequality is found by applying the lower bound of $\underline{Z}_1$ in (3.2.33). From (3.2.60) and (3.2.29) we conclude that $\underline{C}_1 > 1$. Our comparisons of the two control efficiency functions are summarized as follows:

$$\max(1,C_2) < C_1 < 1 + C_2. \tag{3.2.70}$$

We emphasize that the control efficiency functions $\underline{C}_1$ and $\underline{C}_2$ are functions of the two transmission factors $\underline{T}_1$ and $\underline{T}_2$ that are defined everywhere above the threshold where $\underline{T}_1\underline{T}_2 = 1$. The inequalities in (3.2.70) hold throughout the domain of definition of of $\underline{C}_1$ and $\underline{C}_2$, i.e. for all communities above the threshold. Thus we have derived a qualitative result that does not require information about the numerical values of $\underline{T}_1$ and $\underline{T}_2$ for its validity. A given proportionate reduction of $\underline{T}_1$ always leads to a larger reduction of $\underline{X}_1$ (the steady-state expected number of infections per human being) than an equally large proportionate reduction of $\underline{T}_2$, since $\underline{C}_1 > \underline{C}_2$. Furthermore, (3.2.68) shows that the ratio $\underline{C}_1/\underline{C}_2$ is large if $\underline{P}_2$ is large, but that it is close to one if $\underline{P}_2$ is small.

The efficiencies of control of the steady-state infection probability $\underline{P}_1$ among human beings are found to be

$$C_{P_1T_1} = T_1 \frac{\partial P_1}{\partial T_1} = X_1(1-P_1)C_1 \tag{3.2.71}$$

and

$$C_{P_1T_2} = T_2 \frac{\partial P_1}{\partial T_2} = X_1(1-P_1)C_2. \tag{3.2.72}$$

The ratio between $\underline{C}_{P_1T_1}$ and $\underline{C}_{P_1T_2}$ is equal to the ratio $\underline{C}_1/\underline{C}_2$, and the comments made above in the comparison of $\underline{C}_1$ and $\underline{C}_2$ apply also to the comparison of the control efficiency functions $\underline{C}_{P_1T_1}$ and $\underline{C}_{P_1T_2}$.

We emphasize the conceptual importance of the qualitative result that control through reduction of $\underline{T}_1$ is more efficient than control through reduction of $\underline{T}_2$. Similar results can be expected to hold in more realistic models.

A study of the efficiencies of control of the public health factor, the prevalence, the incidence, and the recovery probablity can also be carried out, and a number of qualitative results can be established. We omit details.

3.2.7 The Influence of Superinfection

The results in the preceding two subsections give epidemiological measures and control efficiencies for endemic malaria with superinfection. These results can be compared with the corresponding results for the Ross malaria model in Section 3.1. The comparison gives an indication of the influence of the superinfection hypothesis on the epidemiological indicators of infection level and on the control efficiencies. We shall make some such comparisons under the assumption that all parameters have the same values. This means in particular that the recovery rate r_1 in the Ross model has the same value as the rate r_1 of termination of an individual infection in the model with superinfection. A consequence of our assumption is that the two transmission factors have the same values for the two models. In order to distinguish corresponding quantities in the two models, we shall equip those in the Ross model with the subscript R and those in the superinfection model with the subscript S.

The superinfection hypothesis has the consequence that the recovery rate is lowered. Indeed, recovery of an individual with several infections requires that all the infections in the body of the individual be terminated. The ratio between the recovery rate R_1 of the host and the rate r_1 of recovery per individual infection is found from (2.2.49) to be equal to

$$\frac{R_1}{r_1} = \frac{1 - p_1}{p_1} \ln \frac{1}{1 - p_1} , \tag{3.2.73}$$

where p_1 is the infection probability. Some values of this ratio of recovery rates are given in Table 3.2.2 on the next page.

We proceed to show that in steady state the two infection probabilities P_1 and P_2 and the public health factor H_1 are larger in the model that accounts for superinfection than in the Ross model, and also that the prevalence at each age is larger with superinfection. We have already noted that the lower bound of Z_1 in (3.2.33) means that the steady-state infection probability among mosquitoes is larger under superinfection:

Table 3.2.2 The ratio between the host recovery rate R_1 for the superinfection model and the rate of recovery per individual infection r_1 for various values of the infection probability p_1.

p_1	R_1/r_1
0.01	0.9950
0.1	0.9482
0.3	0.8322
0.5	0.6931
0.7	0.5160
0.9	0.2558
0.99	0.0465

$$P_{2S} > P_{2R}. \tag{3.2.74}$$

In order to show the corresponding inequality between the infection probabilities among human beings, we note from relation (3.2.31), the definition (3.2.19) of the function f, and the expression (3.2.51) for the infection probability P_{1S} that

$$P_{1S} = \frac{1}{T_2} \frac{Z_1}{1-Z_1} . \tag{3.2.75}$$

By applying the lower bound of Z_1 in (3.2.33) and expression (3.1.19) for the human infection probability P_{1R} in the Ross model we find that

$$P_{1S} > P_{1R}. \tag{3.2.76}$$

The steady-state public health factor obeys the relation

$$H_1 = r_1 T_1 P_2 \tag{3.2.77}$$

in both models. It follows therefore from (3.2.74) that

$$H_{1S} > H_{1R}. \tag{3.2.78}$$

The steady-state prevalence at age s can be written

$$Q_R(s) = \frac{H_{1R}}{H_{1R}+r_1}\left(1 - e^{-(H_{1R}+r_1)s}\right) \tag{3.2.79}$$

for the Ross model, and

$$Q_S(s) = 1 - \exp\left(-\frac{H_{1S}}{r_1}\left(1 - e^{-r_1 s}\right)\right) \tag{3.2.80}$$

in the superinfection model. If, under superinfection, the steady-state infection rate were equal to $\underline{H}_{1R}$, then the prevalence at age $\underline{s}$ would be equal to

$$\bar{Q}_S(s) = 1 - \exp\left(-\frac{H_{1R}}{r_1}\left(1 - e^{-r_1 s}\right)\right). \tag{3.2.81}$$

Inequality (3.2.78) between the infection rates allows us to conclude that the resulting prevalence is lower at each positive age $\underline{s}$:

$$\bar{Q}_S(s) < Q_S(s), \qquad s > 0. \tag{3.2.82}$$

We note furthermore that

$$\frac{\bar{Q}'_S(s)}{Q'_R(s)} = \exp\left(\frac{H_{1R}}{r_1}\left(r_1 s - 1 + e^{-r_1 s}\right)\right) > 1,\ s > 0, \tag{3.2.83}$$

since

$$e^{-r_1 s} > 1 - r_1 s,\ s > 0. \tag{3.2.84}$$

We conclude therefore that

$$Q_R(s) < \bar{Q}_S(s)\ , \qquad s > 0, \tag{3.2.85}$$

since

$$Q_R(0) = \bar{Q}_S(0) = 0. \tag{3.2.86}$$

We note that superinfection increases the prevalence in two ways. First of all, if the human infection rates with and without superinfection are the same, then superinfection gives a higher prevalence, as shown by (3.2.85). But the higher human prevalence leads in steady state to a higher mosquito infection probability which in turn leads to a higher human infection rate. This increases the human prevalence further, as shown by inequality (3.2.82).

A comparison of the prevalence versus age in the two models has previously been made by Fine, (1975). The comparison is made under the assumption that the human infection rate and the rate of termination per infection are constant. This amounts to a comparison of host models, not of transmission models. The numerical results presented by Fine illustrate inequality (3.2.85) but not the inequality (3.2.82).

3.3 The Macdonald-Dietz Malaria Model

It was noted in the previous section that the malaria model advanced by Macdonald differed from that of Ross in two respects: Macdonald assumed superinfection in human beings, and he allowed for a latent period in the mosquitoes. The previous section treated a transmission model with superinfection but with mosquito latency ignored. In this section we deal with the full Macdonald-Dietz malaria transmission model, allowing for both superinfection among human beings and latency among mosquitoes.

3.3.1 Model Formulation

We assume that the latent period for each mosquito is constant, and we denote it by $\underline{T}$. In addition, we make use of the seven basic parameters introduced in the previous section.

The model takes the form of a collection of $\underline{N}_1$ Markov chains and three deterministic functions as follows:

$$\{X_1^{(1)}(t),\ X_1^{(2)}(t),\ldots,X_1^{(N_1)}(t),\ S(t),\ L(t),\ I(t)\},\ t \geq 0.$$

Here, $\underline{X}_1^{(k)}(\underline{t})$ stands for the number of infections in human host number $\underline{k}$ at time $\underline{t}$, while $\underline{S}(\underline{t})$, $\underline{L}(\underline{t})$, $\underline{I}(\underline{t})$ represent the number of susceptible, latent, and infected mosquitoes, respectively, at time $\underline{t}$. The values of these three functions satisfy the relation $\underline{S}(\underline{t}) + \underline{L}(\underline{t}) + \underline{I}(\underline{t}) = \underline{N}_2$, $\underline{t} \geq -\underline{T}$. The reason for requiring this relation to hold for $\underline{t} \geq -\underline{T}$ and not only for $\underline{t} \geq 0$ is that initial values must be specified over the whole interval $\left[-\underline{T}, 0\right]$. The hypotheses for the Markov chains $\underline{X}_1^{(k)}(\underline{t})$ are described in Section 2.2, and those for the mosquito phase in Section 2.7. By using equations (2.2.32) and (2.7.7)-(2.7.9) we are led to the following system of equations:

$$x_1' = h_1(t) - r_1x_1, \tag{3.3.1}$$

$$S' = r_2N_2 - h_2(t)S - r_2S, \tag{3.3.2}$$

$$L' = h_2(t)S - e^{-r_2T}h_2(t-T)S(t-T) - r_2L, \tag{3.3.3}$$

$$I' = e^{-r_2T}h_2(t-T)S(t-T) - r_2I. \tag{3.3.4}$$

Here, the argument of the state variables $\underline{x}_1$, $\underline{S}$, $\underline{L}$, $\underline{I}$ is either $\underline{t}$ or $\underline{t} - \underline{T}$. The argument is omitted from the notation when it is equal to $\underline{t}$. The function value $\underline{x}_1(\underline{t})$ stands for the expected number of infections per human host at time $\underline{t}$. The infection rates $\underline{h}_1(\underline{t})$ and $\underline{h}_2(\underline{t})$ are found to be

$$h_1(t) = \frac{a_2b_2}{N_1} I(t), \tag{3.3.5}$$

$$h_2(t) = a_2b_1(1 - e^{-x_1(t)}). \tag{3.3.6}$$

Only one hybrid assumption is necessary in the derivation of these expressions, since the number of infected mosquitoes $\underline{I}(\underline{t})$ is already a deterministic quantity. The hybrid assumption used is that the number of infected human beings is replaced by its expected value $\underline{N}_1(1 - \underline{e}^{-x_1(t)})$. We note that the equation (3.3.3) for the number of mosquitoes in the latent state is not needed for the further analysis. By inserting the expressions (3.3.5), (3.3.6) for the infection rates into the equation (3.3.1), (3.3.2), (3.3.4), we find that the functions $\underline{x}$, $\underline{S}$, and $\underline{I}$ satisfy the following system of differential - difference equations:

$$x_1' = \frac{a_2 b_2}{N_1} I - r_1 x_1, \tag{3.3.7}$$

$$S' = r_2 N_2 - a_2 b_1 (1 - e^{-x_1}) S - r_2 S, \tag{3.3.8}$$

$$I' = e^{-r_2 T} a_2 b_1 (1 - e^{-x_1(t-T)}) S(t-T) - r_2 I. \tag{3.3.9}$$

3.3.2 Quasi-Dimensional Analysis

The next step in the analysis is to introduce suitable state variables that are free of quasi-dimesion, and to derive a system of equations for them. Our goal is to reduce as much as possible the number of parameters necessary to describe the steady-state behaviour of the solution to the system of equations. In order to do this we analyze the critical points $(\bar{\underline{x}}_1, \bar{\underline{S}}, \bar{\underline{I}})$ of equations (3.3.7)-(3.3.9). By putting the right-hand sides equal to zero we find after elimination of $\bar{\underline{S}}$ and $\bar{\underline{I}}$ that $\bar{\underline{x}}_1$ satisfies the relation

$$\bar{x}_1 = \frac{a_2 b_1}{r_2} \left(1 - e^{-\bar{x}_1}\right) \left(\frac{a_2 b_2 N_2}{r_1 N_1} e^{-r_2 T} - \bar{x}_1 \right). \tag{3.3.10}$$

We note that the first factor of the right-hand side of this expression is equal to $\underline{T}_2$ as defined in (3.2.9). We introduce the same notation here. Furthermore, we put $\underline{T}_1$ equal to the first term of the third factor of the right-hand side. Thus we define

$$T_1 = \frac{a_2 b_2 N_2}{r_1 N_1} e^{-r_2 T} \tag{3.3.11}$$

and

$$T_2 = \frac{a_2 b_1}{r_2} . \tag{3.3.12}$$

A comparison with (3.2.8) shows that latency reduces the transmission factor $\underline{T}_1$ by the factor $\underline{e}^{-r_2T}$, which equals the proportion of mosquitoes that survive the latent state in steady state. The quasi-dimensions of $\underline{T}_1$ and $\underline{T}_2$ are clearly the same as in Section 3.2.

We proceed to introduce three variables free of quasi-dimension by putting

$$z_1 = x_1/T_1, \tag{3.3.13}$$

$$z_2 = S/N_2, \tag{3.3.14}$$

$$z_3 = I/N_2. \tag{3.3.15}$$

We find then from equations (3.3.7) - (3.3.9) that these new variables satisfy the following system of differential-difference equations:

$$z_1' = r_1\left(e^{r_2T}z_3 - z_1\right), \tag{3.3.16}$$

$$z_2' = r_2\left(1 - z_2 - T_2\left(1-e^{-T_1z_1}\right)z_2\right), \tag{3.3.17}$$

$$z_3' = r_2\left(T_2e^{-r_2T}\left(1-e^{-T_1z_1(t-T)}\right)z_2(t-T) - z_3\right). \tag{3.3.18}$$

These equations are to be satisfied for $\underline{t} \geq 0$. In order for a unique solution to exist, it is necessary to specify values of all three functions throughout the initial interval $-\underline{T} \leq \underline{t} \leq 0$. The specifications take the form

$$z_i(t) = g_i(t), \quad -T \leq t \leq 0, \quad i=1,2,3. \tag{3.3.19}$$

The three functions $\underline{g}_i(\underline{t})$, $\underline{i} = 1,2,3$, are given continuous non-negative functions with $g_2(\underline{t}) + g_3(\underline{t}) \leq 1$, $-\underline{T} \leq \underline{t} \leq 0$. The form of the equations shows that it is unnecessary to specify $g_3(\underline{t})$ throughout the initial interval; it suffices to specify its value at $\underline{t} = 0$. For notational convenience in the following analysis we shall however assume that g_3 is specified as above. We introduce $\underline{z}(\underline{t})$ to denote the column vector with components $\underline{z}_i(\underline{t})$,

$\underline{i} = 1,2,3$, $\underline{t} \geq -\underline{T}$. We define $\underline{z}(\underline{t})$, $\underline{t} \geq -\underline{T}$, to be a solution of the initial-value problem (3.3.16) - (3.3.19) if it satisfies the system of equation (3.3.16) - (3.3.18) for $\underline{t} \geq 0$, the initial conditions (3.3.19) on the initial interval $[-\underline{T},0]$, and is continuous for $\underline{t} \geq -\underline{T}$.

3.3.3. Model Analysis

We proceed to find the equilibrium solutions of this system of equations and to analyze their local stability properties. To agree with the notation used in Appendix IV, we write the system of equations in the vector form

$$z'(t) = f(z(t), z(t-T)). \tag{3.3.20}$$

Any equilibrium solution $\bar{\underline{z}}$ satisfies the equation $\underline{f}(\bar{\underline{z}}, \bar{\underline{z}}) = 0$. We readily find that the components of the vector $\bar{\underline{z}}$ satisfy

$$\bar{z}_2 = 1 - \bar{z}_1 \tag{3.3.21}$$

and

$$\bar{z}_3 = e^{-r_2 T} \bar{z}_1 , \tag{3.3.22}$$

where $\bar{\underline{z}}_1$ is a solution of the equation

$$\bar{z}_1 = T_2(1 - e^{-T_1 \bar{z}_1})(1 - \bar{z}_1). \tag{3.3.23}$$

We note that this equation is identical to equation (3.2.17) for the $\underline{z}_1$-component of any critical point of the system of equations (3.2.12), (3.2.13). Thus, the results of Section 3.2 show that the column vector $\bar{\underline{z}}^0 = (0,1,0)^T$ is an equilibrium solution of (3.3.20) for all parameter values, and that in addition the column vector $\bar{\underline{z}}^1 = (\underline{Z}_1, 1-\underline{Z}_1, \underline{e}^{-r_2 T} \underline{Z}_1)^T$ is a critical point defined for $\underline{T}_1 \underline{T}_2 \geq 1$. Here, $\underline{Z}_1$ denotes the function value $\underline{Z}_1(\underline{T}_1,\underline{T}_2)$ that satisfies relation (3.2.28). The relation $\underline{T}_1 \underline{T}_2 = 1$ will be referred to as the threshold. We note that it is formally identical to the thresholds for the two malaria models in Sections 3.1 and 3.2. The point $\bar{\underline{z}}^0$ corresponds to the absence of infection in the community, while the point $\bar{\underline{z}}^1$ above the threshold corresponds to an endemic infection level. The two critical points coincide on the threshold.

By denoting the two vector arguments of the function $\underline{f}$ in (3.3.20) by $\underline{u}$ and $\underline{v}$, we find from (3.3.16) - (3.3.18) that $\underline{f}(\underline{u},\underline{v})$ can be written as follows:

$$f(u,v) = \begin{pmatrix} -r_1u_1 + r_1e^{r_2T}\, u_3 \\ r_2 - r_2(1+T_2)u_2 + r_2T_2e^{-T_1u_1}\, u_2 \\ -r_2u_3 + r_2T_2e^{-r_2T}\, v_2 - r_2T_2e^{-r_2T-T_1v_1}\, v_2 \end{pmatrix} \qquad (3.3.24)$$

The two Jacobian matrices $\frac{\partial f}{\partial u}$ and $\frac{\partial f}{\partial v}$ are therefore

$$\frac{\partial f}{\partial u}(u,v) = \begin{pmatrix} -r_1 & 0 & r_1e^{r_2T} \\ -r_2T_1T_2e^{-T_1u_1}u_2 & -r_2-r_2T_2\left(1-e^{-T_1u_1}\right) & 0 \\ 0 & 0 & -r_2 \end{pmatrix}$$

(3.3.25)

and

$$\frac{\partial f}{\partial v}(u,v) = \begin{pmatrix} 0 & 0 & 0 \\ 0 & 0 & 0 \\ r_2T_1T_2e^{-r_2T-T_1v_1}v_2 & r_2T_2e^{-r_2T}\left(1-e^{-T_1v_1}\right) & 0 \end{pmatrix}.$$

(3.3.26)

We first evaluate these matrices when both arguments of $\underline{f}$ are equal to $\bar{\underline{z}}^0$, the critical point corresponding to no infection in the community. We get

$$B_0 = -\frac{\partial f}{\partial u}(\bar{z}^0,\bar{z}^0) = \begin{pmatrix} r_1 & 0 & -r_1 e^{r_2 T} \\ r_2 T_1 T_2 & r_2 & 0 \\ 0 & 0 & r_2 \end{pmatrix} \qquad (3.3.27)$$

and

$$B_1 = -\frac{\partial f}{\partial v}(\bar{z}^0,\bar{z}^0) = \begin{pmatrix} 0 & 0 & 0 \\ 0 & 0 & 0 \\ -r_2 T_1 T_2 e^{-r_2 T} & 0 & 0 \end{pmatrix}. \qquad (3.3.28)$$

The characteristic matrix function $\underline{H}(\underline{s}) = \underline{I}\underline{s} + \underline{B}_0 + \underline{B}_1 e^{-Ts}$ has the determinant

$$h(s) = (s+r_2)e^{-Ts}\, k(s) \;, \qquad (3.3.29)$$

where

$$k(s) = (s^2 + ps + q)e^{Ts} - r \qquad (3.3.30)$$

with

$$p = r_1 + r_2,$$

$$q = r_1 r_2, \qquad (3.3.31)$$

and

$$r = r_1 r_2 T_1 T_2.$$

It is readily verified that the conditions in (AIV.9) are satisfied. It follows therefore that all the roots of the characteristic equation $\underline{h}(\underline{s}) = 0$ have negative real parts if $\underline{r} < \underline{q}$, i.e. if $\underline{T}_1\underline{T}_2 < 1$, and that at least one root has positive real part if $\underline{T}_1\underline{T}_2 > 1$. We conclude that the equilibrium solution $\underline{\bar{z}}^0$, which corresponds to an infection-free community, is asymptotically stable for $\underline{T}_1\underline{T}_2 < 1$, and unstable for $\underline{T}_1\underline{T}_2 > 1$.

We proceed to evaluate the two Jacobian matrices when both arguments of $\underline{f}$ are equal to $\underline{\bar{z}}^1$, the critical point that corresponds to an endemic infection level above the threshold. By using relation (3.3.23) to achieve some simplifications, we get

$$B_0 = -\frac{\partial f}{\partial u}(\bar{z}^1,\bar{z}^1) = \begin{pmatrix} r_1 & 0 & -r_1 e^{r_2 T} \\ r_2 T_1 (T_2-(T_2+1)Z_1) & \frac{r_2}{1-Z_1} & 0 \\ 0 & 0 & r_2 \end{pmatrix} \tag{3.3.32}$$

and

$$B_1 = -\frac{\partial f}{\partial v}(\bar{z}^1,\bar{z}^1) = \begin{pmatrix} 0 & 0 & 0 \\ 0 & 0 & 0 \\ -r_2 T_1 e^{-r_2 T}(T_2-(T_2+1)Z_1) & -\frac{r_2 e^{-r_2 T} Z_1}{1-Z_1} & 0 \end{pmatrix} \tag{3.3.33}$$

The characteristic matrix function $\underline{H}(\underline{s})$ has the determinant

$$h(s) = (s + r_2)e^{-Ts}\, k(s), \tag{3.3.34}$$

where

$$k(s) = (s^2 + ps + q)e^{Ts} - r, \tag{3.3.35}$$

with

$$p = r_1 + \frac{r_2}{1-Z_1},$$

$$q = \frac{r_1 r_2}{1-Z_1}, \tag{3.3.36}$$

and

$$r = r_1 r_2 T_1 (T_2 - (T_2+1)Z_1).$$

It is readily verified that the conditions in (AIV.9) are satisfied. It follows therefore that all the roots of the characteristic equation $\underline{h}(\underline{s}) = 0$ have negative real parts if $\underline{r} < \underline{q}$, i.e. if

$$T_1(1-Z_1)(T_2 - (T_2+1)Z_1) < 1 . \tag{3.3.37}$$

The upper bound of $\underline{Z}_1$ in (3.2.33) shows that the left-hand side of this inequality is always positive. An application of the lower bound in (3.2.33), which is a strict lower bound everywhere above the threshold, shows that the left-hand side of the inequality is bounded above as follows:

$$T_1(1-Z_1)(T_2-(T_2+1)Z_1) < \frac{T_1+1}{T_1T_2+T_1} < 1, \quad T_1T_2 > 1. \tag{3.3.38}$$

We conclude that the equilibrium solution $\underline{z}^1$ is asymptotically stable for $\underline{T}_1\underline{T}_2 > 1$, i.e. everywhere above the threshold.

This completes our mathematical analysis of the model. We note that we have only established the local stability of the equilibrium solutions.

The epidemiological interpretation of the results is straightforward and is formally identical to the one carried out in Section 3.2. The main conclusion to be drawn from this model is that a constant latent period in the mosquitoes is accounted for by its effect on the transmission factor $\underline{T}_1$ as seen in (3.3.11): Latency reduces the transmission factor $\underline{T}_1$ by a factor which measures the proportion of mosquitoes that survive the latent state in steady state.

Macdonald (1952) has introduced the concept of basic reproduction rate in malaria and shown its importance in judging the effect of various eradication measures. The basic reproduction rate can be interpreted as the average number of secondary cases arising from a single primary case in a large population of susceptible hosts. An endemic infection level can exist if and only if the basic reproduction rate is larger than one. It turns out that the basic reproduction rate is equal to the product of the two transmission factors $\underline{T}_1$ and $\underline{T}_2$.

Macdonald denotes the basic reproduction rate by $\underline{z}_0$, and shows that is equal to (with his notation) the classical expression

$$z_0 = \frac{ma^2b\,p^n}{-r\log p}\,. \tag{3.3.39}$$

To see that this expression is equal to $\underline{T}_1\underline{T}_2$, we interpret Macdonald's notation in ours. Indeed, $\underline{m} = \underline{N}_2/\underline{N}_1$ is the number of female mosquitoes per human being, $\underline{a} = \underline{a}_2$ is the mosquito man biting rate, $\underline{b} = \underline{b}_2$ is the mosquito infectivity, $\underline{r} = \underline{r}_1$ is the rate of recovery per infection in human beings, $\underline{p} = \exp(-\underline{r}_2)$ is the probability for a mosquito to survive one day, and $\underline{n} = \underline{T}$ is the length of the mosquito latent period. It follows that $\log \underline{p} = -\underline{r}_2$. Moreover, Macdonald assumes that the human infectivity is $\underline{b}_1 = 1$. Insertion of these expressions into (3.3.39) shows that

$$z_0 = T_1 T_2 \ . \tag{3.3.40}$$

The analysis of epidemiological measures in the endemic case and of control efficiencies is identical to the corresponding analysis in Section 3.2, since the effect of latency in the mosquito population is accounted for by a redefinition of the transmission factor $\underline{T}_1$.

CHAPTER 4. TRANSMISSION MODELS FOR HERMAPHRODITIC HELMINTHIASIS

The life cycle of hermaphroditic helminths is very similar to that of schistosomes. The main difference from a transmission standpoint is that eggs are laid by all sexually mature parasites in hermaphroditic helminthiasis, while only mated female parasites lay eggs in schistosomiasis. This has the consequence that the infectivity of the definitive host population is determined by the total number of mature parasites in hermaphroditic helminthiasis and by the total number of mated female parasites in schistosomiasis. This consequence is of direct imortance for the formulation of transmission models for the two infections. In both cases, it is necessary to account for superinfection of definitive hosts.

In this chapter we formulate and study two transmission models for hermaphroditic helminthiasis. The first one ("the basic model") ignores immune reactions of human beings, while the second one accounts for concomitant immunity.

If mating between male and female schistosomes were ignored, then all female schistosomes would lay eggs. A schistosomiasis model that ignores mating would therefore be essentially equivalent to a model for hermaphroditic helminthiasis.

4.1 The Basic Model

In the basic model for transmission of hermaphroditic helminthiasis, the superinfection process of Section 2.2 serves as model for the number of parasites in each human host. We assume that the size of the snail population is constant and that the snails are subject to birth and death. Thus, the second model of Section 2.6 serves as model for the number of infected snails. The model discussed here differs from the basic schistosomiasis model of Section 5.1 only by ignoring the effect of mating between male and female parasites. The model has been presented by Nåsell and Hirsch (1972).

4.1.1 Model Formulation

We consider a closed ecological community as a possible site for transmission and we assume that the population of human beings is constant and that there is a constant number of snails. Further assumptions are implicit in the following definitions of eight basic parameters:

$\underline{N}_1$ = the number of human hosts,

μ_1 = the death rate per sexually mature parasite,

λ_1 = the egg-laying rate = the number of eggs laid per unit time by each sexually mature parasite,

π_1 = the probability for each cercaria to infect a given human being,

$\underline{N}_2$ = the number of snails,

μ_2 = the death rate per snail,

λ_2 = the rate of cercarial shedding = the number of cercariae emitted per unit time by each infected snail,

π_2 = the probability for each egg to infect a given snail.

The model takes the explicit form of a collection of Markov chains as follows:

$$\{X_1^{(1)}(t),\ldots, X_1^{(N_1)}(t),\ X_2(t)\}\ ,\quad t \geq 0.$$

Thus, to each human host $\underline{k}$ there corresponds a Markov chain $\underline{X}_1^{(k)}$ that gives the number of sexually mature parasites in that host. The state space of each of these Markov chains is the set of all nonnegative integers. The Markov chain $\underline{X}_2$ gives the number of infected snails. The state space of this Markov chain is the set of all nonnegative integers less than or equal to $\underline{N}_2$, the total number of snails.

Each $\underline{X}_1^{(k)}$ is a superinfection process (Section 2.2) with infection rate $\underline{h}_1(\underline{t})$ and death rate per parasite μ_1. The process $\underline{X}_2$ models infection in the population of snails according to the second model of Section 2.6 with infection rate per uninfected snail $\underline{h}_2(\underline{t})$ and death rate per snail μ_2. We assume that initially the $\underline{N}_1$ random variables $\underline{X}_1^{(k)}(0)$ have Poisson distribution with the same parameter $\underline{x}_{10}$, and that $\underline{X}_2(0)$ has a binomial distribution with parameters $\underline{N}_2$ and $\underline{p}_{20}$. It follows from these initial assumptions that the number of parasites per human host has a Poisson distribution for each $\underline{t} \geq 0$ with time-dependent parameter $\underline{x}_1(\underline{t})$ independent of the individual host, and that the infection probability $\underline{p}_2(\underline{t})$ is the same for all snails. It follows furthermore from (2.2.32) that the expected number of parasites per human host $\underline{x}_1$ satisfies the differential equation

$$x_1' = h_1(t) - \mu_1 x_1. \tag{4.1.1}$$

The infection probability for each snail $\underline{p}_2$ is found from (2.1.23) to satisfy the differential equation

$$p_2' = h_2(t)(1 - p_2) - \mu_2 p_2. \tag{4.1.2}$$

In order to complete the model specification, we express the infection rates $\underline{h}_1(\underline{t})$ and $\underline{h}_2(\underline{t})$ in terms of parameters and state variables. The infection rate $\underline{h}_1(\underline{t})$ is the rate at which cercariae infect each given human host at time $\underline{t}$. It is the product of the total rate at which cercariae are emitted into the environment and the probability π_1 for each cercaria to infect a given human host. Each infected snail emits cercariae at the rate λ_2. The total rate of production of cercariae is equal to λ_2 multiplied by the total number of infected snails at time $\underline{t}$. The number of infected snails is, in the model, a stochastically varying quantity. One of the hybrid assumptions consists in replacing this quantity by its expected value $\underline{N}_2\underline{p}_2(\underline{t})$. Thus we get

$$h_1(t) = \pi_1 \lambda_2 N_2 p_2(t). \tag{4.1.3}$$

In a similar way we find that the infection rate $\underline{h}_2(\underline{t})$ per susceptible snail is the product of the total rate of egg production and the probability π_2 for each egg to infect a given snail.

Each mature parasite lays eggs at the rate λ_1. The total rate of egg production in the community is equal to λ_1 multiplied by the total number of parasites at time $\underline{t}$. The total number of parasites is in the model represented by a stochastic quantity. The second hybrid assumption consists in replacing this quantity by its expected value $\underline{N}_1\underline{x}_1(\underline{t})$. Thus we get

$$h_2(t) = \pi_2\lambda_1N_1x_1(t). \tag{4.1.4}$$

By inserting these expressions for the infection rates into equations (4.1.1), (4.1.2), we are led to the following system of differential equations for the state variables $\underline{x}_1$, $\underline{p}_2$:

$$x_1' = \pi_1\lambda_2N_2p_2 - \mu_1x_1, \tag{4.1.5}$$

$$p_2' = \pi_2\lambda_1N_1x_1(1 - p_2) - \mu_2p_2. \tag{4.1.6}$$

The initial value for $\underline{x}_1$ is nonnegative, $0 \leq \underline{x}_{10}$, and the initial value for $\underline{p}_2$ is constrained to lie in the unit interval, $0 \leq \underline{p}_{20} \leq 1$.

4.1.2 Quasi-Dimensional Analysis

We proceed with a quasi-dimensional analysis of equations (4.1.5), (4.1.6). Our goal is to reduce as much as possible the number of parameters needed to determine the steady-state solutions of the equations.

As fundamental dimensional unit we use time $\underline{T}$, and as fundamental quasi-dimensional units we use "human beings" $\underline{H}$, "parasites" $\underline{P}$, "snails" $\underline{S}$, "eggs" $\underline{E}$, and "cercariae" $\underline{C}$. We then find that the number of human beings $\underline{N}_1$ has quasi-dimension $\underline{H}$, and that the number of snails $\underline{N}_2$ has quasi-dimension $\underline{S}$. Furthermore, the death rates μ_1 and μ_2 both have quasi-dimensions $\underline{T}^{-1}$. The egg-laying rate λ_1 measures the number of eggs laid per unit time by each parasite. Its quasi-dimension is therefore $\underline{E}\ \underline{P}^{-1}\underline{T}^{-1}$. The rate of cercarial shedding λ_2 measures the number of cercariae emitted in unit time by one infected snail; its quasi-dimension is $\underline{C}\ \underline{S}^{-1}\underline{T}^{-1}$.

The parameter π_1 gives the probability for each cercaria to infect a given human being. One might therefore think that π_1 is free of quasi-dimension. However, it is important to recognize that a cercaria that successfully infects a human being develops into a new parasite. Therefore, the probability for a cercaria to infect any human being has quasi-dimension $\underline{P}\ \underline{C}^{-1}$. It follows that the probability π_1 for a cercaria to infect a given human being has quasi-dimension $\underline{P}\ \underline{C}^{-1}\underline{H}^{-1}$. Similarly, the probability for an egg to infect any snail has quasi-dimension $\underline{S}\ \underline{E}^{-1}$. It follows that the probability π_2 for an egg to infect a given snail has quasi-dimension $\underline{E}^{-1}$. The state variable $\underline{x}_1$ gives the expected number of parasites per human host; its quasi-dimension is $\underline{P}\ \underline{H}^{-1}$. The state variable $\underline{p}_2$ gives the probability for any snail to be infected and is free of quasi-dimension. The time variable $\underline{t}$ has quasi-dimension $\underline{T}$. The results of our discussion are summarized in Table 4.1 on the next page.

By using expressions (4.1.3) and (4.1.4) for the infection rates $\underline{h}_1(\underline{t})$ and $\underline{h}_2(\underline{t})$, we find their quasi-dimensions to be $\underline{P}\ \underline{H}^{-1}\underline{T}^{-1}$ and $\underline{T}^{-1}$, respectively.

The next step in the quasi-dimensional analysis is to find a parameter combination that has the same quasi-dimension as the state variable $\underline{x}_1$, namely $\underline{P}\ \underline{H}^{-1}$. (We note that the state variable $\underline{p}_2$ is already free of quasi-dimension).

We illustrate a different method for finding such a parameter combination than the methods used for the malaria models. We denote the parameter combination we are searching for by $\underline{T}_1$. It has the form of a product of powers of the eight basic parameters that have been introducced. It can therefore be written

$$T_1 = N_1^{a_1} N_2^{a_2} \mu_1^{a_3} \mu_2^{a_4} \lambda_1^{a_5} \lambda_2^{a_6} \pi_1^{a_7} \pi_2^{a_8} , \qquad (4.1.7)$$

where $\underline{a}_1,\ldots,\underline{a}_8$ are unknown powers. By requiring the quasi-dimensions of both sides of this equation to be equal we are led to the following relation:

Table 4.1 Quasi-dimensions of parameters and state variables in the basic model for hermaphroditic helminthiasis.

Parameter or state variable	Quasi-dimension
N_1	H
N_2	S
μ_1, μ_2	T^{-1}
λ_1	$E\ P^{-1}T^{-1}$
λ_2	$C\ S^{-1}T^{-1}$
π_1	$P\ C^{-1}H^{-1}$
π_2	E^{-1}
x_1	$P\ H^{-1}$
p_2	1
t	T
h_1	$P\ H^{-1}T^{-1}$
h_2	T^{-1}
T_1	$P\ H^{-1}$
T_2	$H\ P^{-1}$

$$PH^{-1} = H^{a_1-a_7}\,S^{a_2-a_6}\,E^{a_5-a_8}\,P^{a_7-a_5}\,C^{a_6-a_7}\,T^{-a_3-a_4-a_5-a_6}. \tag{4.1.8}$$

By equating the powers of the various fundamental quasi-dimensional units, we get the following relations between the unknown powers $\underline{a}_1,\ldots,\underline{a}_8$:

$$\begin{aligned} a_1 - a_7 &= -1, \\ a_2 - a_6 &= 0, \\ a_5 - a_8 &= 0, \\ a_7 - a_5 &= 1, \\ a_6 - a_7 &= 0, \\ a_3 + a_4 + a_5 + a_6 &= 0. \end{aligned} \tag{4.1.9}$$

Thus, we have six equations for the eight unknown powers. These equations can be used to express any six of the eight powers in terms of the remaining two. A little exploration reveals that it is convenient to solve the equations for $\underline{a}_2$, $\underline{a}_3$, $\underline{a}_5$ - $\underline{a}_8$ in terms of $\underline{a}_1$ and $\underline{a}_4$. The solution has the following form:

$$\begin{aligned} a_2 &= a_1 + 1, \\ a_3 &= -2a_1 - a_4 - 1, \\ a_5 &= a_1, \\ a_6 &= a_1 + 1, \\ a_7 &= a_1 + 1, \\ a_8 &= a_1. \end{aligned} \tag{4.1.10}$$

Insertion of these relations into (4.1.7) shows that $\underline{T}_1$ can be written

$$T_1 = \left[\frac{N_1 N_2 \lambda_1 \lambda_2 \pi_1 \pi_2}{\mu_1^2}\right]^{a_1} \left[\frac{\mu_2}{\mu_1}\right]^{a_4} \frac{N_2 \lambda_2 \pi_1}{\mu_1} . \tag{4.1.11}$$

Our development shows that $\underline{T}_1$ defined by this relation for any choice of $\underline{a}_1$ and $\underline{a}_4$ has the desired quasi-dimension $\underline{P}^{-1}\underline{H}$. This fact indicates the non-uniqueness of the resulting parametrization. The simplest expression for $\underline{T}_1$ is obtained by putting $\underline{a}_1 = \underline{a}_4 = 0$. For this choice of $\underline{a}_1$ and $\underline{a}_4$ we get the following expression for $\underline{T}_1$ from (4.1.11):

$$T_1 = \frac{\pi_1 \lambda_2 N_2}{\mu_1} . \tag{4.1.12}$$

We now introduce a new state variable $\underline{z}_1$ free of quasi-dimension by putting

$$z_1 = \frac{x_1}{T_1} . \tag{4.1.13}$$

For notational convenience we also introduce

$$z_2 = p_2. \tag{4.1.14}$$

By using equations (4.1.5), (4.1.6) we are led to the following system of equations for z_1 and z_2:

$$z_1' = \mu_1(z_2 - z_1), \tag{4.1.15}$$

$$z_2' = \mu_2(T_1 T_2 z_1(1-z_2) - z_2), \tag{4.1.16}$$

where

$$T_2 = \frac{\pi_2 \lambda_1 N_1}{\mu_2} . \tag{4.1.17}$$

As in previous sections we shall refer to T_1 and T_2 as transmission factors. Relation (4.1.13) shows that T_1 serves as an intrinsic reference measure of the number of parasites per human host. The steady-state solutions of equations (4.1.15), (4.1.16) are clearly determined by only one parameter, namely the product $T_1 T_2$ of the two transmission factors. However, the state variables z_1 and z_2 of the system of equations (4.1.15), (4.1.16) are not both epidemiologically meaningful; relation (4.1.13) is needed to relate z_1 to the expected number of parasites per human host x_1. In order to relate the model predictions to epidemiologically meaningful quantities we shall retain the parameter space where the two transmission factors T_1 and T_2 take their values.

Initial values for the two functions z_1 and z_2 are given at $t = 0$ by $z_{10} = x_{10}/T_1$ and $z_{20} = p_{20}$, respectively. We note that the initial point belongs to the rectangle R defined by

$$R = \{(z_1, z_2) : 0 \leq z_1 \leq \hat{z}_1,\ 0 \leq z_2 \leq 1\}, \tag{4.1.18}$$

where

$$\hat{z}_1 = \max(1,\ z_{10}). \tag{4.1.19}$$

We proceed in the next subsection to analyze the initial-value problem posed by the equations (4.1.15), (4.1.16) with initial point in the rectangle R. In particular, we shall show that R is positively invariant and that every solution approaches a critical point as time approaches infinity.

4.1.3 Model Analysis

We begin the analysis of the initial value problem posed by the system of equations (4.1.15), (4.1.16) with initial point in the rectangle $\underline{R}$ defined in (4.1.18) by identifying the isoclines where the two derivatives are equal to zero. We note first that $\underline{z}_1' = 0$ on the isocline

$$z_2 = F_1(z_1) = z_1, \tag{4.1.20}$$

and that $\underline{z}_2' = 0$ on the isocline

$$z_2 = F_2(z_1) = \frac{T_1 T_2 z_1}{1 + T_1 T_2 z_1} . \tag{4.1.21}$$

Critical points of the system of equations correspond to equilibrium solutions and are found as intersections of the isoclines. We denote any critical point by $(\bar{\underline{z}}_1, \bar{\underline{z}}_2)$. From the equations for the isoclines we find that the $\underline{z}_1$-component of any critical point satisfies

$$\bar{z}_1 = \frac{T_1 T_2 \bar{z}_1}{1 + T_1 T_2 \bar{z}_1} \tag{4.1.22}$$

and that

$$\bar{z}_2 = \bar{z}_1 . \tag{4.1.23}$$

Equation (4.1.22) has two solutions, namely

$$\bar{z}_1 = \begin{cases} 0, \\ 1 - \dfrac{1}{T_1 T_2} . \end{cases} \tag{4.1.24}$$

The second of these solutions is nonnegative only for $\underline{T}_1 \underline{T}_2 \geq 1$. Since we show below that the rectangle $\underline{R}$ is positively invariant we can conclude that the critical point corresponding to the second solution of equation (4.1.22) can be realized as an equilibrium solution only for $\underline{T}_1 \underline{T}_2 \geq 1$.

We proceed to analyze the local stability of the critical points. By writing the system of equations (4.1.15), (4.1.16) in the vector form $\underline{z}' = \underline{f}(\underline{z})$, we find that the Jacobian matrix of $\underline{f}$ can be written

$$\frac{\partial f}{\partial z}(z) = \begin{pmatrix} -\mu_1 & \mu_1 \\ \mu_2 T_1 T_2 (1-z_2) & -\mu_2(1+T_1 T_2 z_1) \end{pmatrix}. \qquad (4.1.25)$$

At the origin we find that

$$\frac{\partial f}{\partial z}(0) = \begin{pmatrix} -\mu_1 & \mu_1 \\ \mu_2 T_1 T_2 & -\mu_2 \end{pmatrix}. \qquad (4.1.26)$$

The eigenvalues of this matrix satisfy the equation

$$\lambda^2 + (\mu_1+\mu_2)\lambda + \mu_1 \mu_2 (1 - T_1 T_2) = 0. \qquad (4.1.27)$$

It follows from (AIV.17) that the origin is asymptotically stable if $\underline{T}_1 \underline{T}_2 < 1$, and that the origin is unstable if $\underline{T}_1 \underline{T}_2 > 1$. We use $\underline{\bar{z}}^1$ to denote the critical point for which both components are equal to $1 - 1/\underline{T}_1 \underline{T}_2$. The Jacobian matrix evaluated at $\underline{\bar{z}}^1$ is found to be

$$\frac{\partial f}{\partial z}(\bar{z}^1) = \begin{pmatrix} -\mu_1 & \mu_1 \\ \mu_2 & -\mu_2 T_1 T_2 \end{pmatrix}. \qquad (4.1.28)$$

The eigenvalues of this matrix satisfy the equation

$$\lambda^2 + (\mu_1 + \mu_2 T_1 T_2) + \mu_1 \mu_2 (T_1 T_2 - 1) = 0. \qquad (4.1.29)$$

It follows from (AIV.17) that the equilibrium solution $\underline{\bar{z}}^1$ is asymptotically stable for $\underline{T}_1 \underline{T}_2 > 1$.

Phase portraits for the system of equations (4.1.15), (4.1.16) are shown in Figures 4.1A and 4.1B for two different cases. Figure 4.1A illustrates the case when $\underline{T}_1 \underline{T}_2 \leq 1$ with the origin as the only critical point in the rectangle $\underline{R}$. Figure 4.1B applies when $\underline{T}_1\underline{T}_2 > 1$ with two distinct critical points in the rectangle $\underline{R}$. Arguments that have been applied before in the study of phase portraits can be used to show that the rectangle $\underline{R}$ is positively invariant for all positive values of the transmission factors $\underline{T}_1$ and $\underline{T}_2$. Furthermore, the origin is globally stable if $\underline{T}_1\underline{T}_2 \leq 1$, i.e. the whole rectangle $\underline{R}$ belongs to its domain of attraction. If $\underline{T}_1\underline{T}_2 > 1$, then the critical point with both coordinates equal to $1 - 1/\underline{T}_1\underline{T}_2$ has all of the rectangle $\underline{R}$ with the exception of the origin as its domain of attraction. In that case, the origin is an unstable equilibrium solution since any initial point in $\underline{R}$ different from the origin gives rise to a solution that approaches the second critical point $\underline{\bar{z}}^1$ as $\underline{t} \to \infty$.

Our results allow us to conclude that there is a threshold function in parameter space, defined by the relation $\underline{T}_1\underline{T}_2 = 1$. Any community for which the pair of transmission factors $\underline{T}_1, \underline{T}_2$ lies above the threshold can support an endemic infection. Any amount of infection brought into such a community will establish itself at an endemic level determined by the two transmission factors. A community for which the transmission factors lie on or below the threshold function cannot support an endemic infection. Eradication of an established infection can be achieved by reducing $\underline{T}_1\underline{T}_2$ to a value smaller than or equal to one. The product $\underline{T}_1\underline{T}_2$ of the two transmission factors is a measure of the amount of effort required to achieve eradication. It is termed the eradication effort.

4.1.4 Epidemiological Measures in the Endemic Case

We consider a community above threshold, so that $\underline{T}_1\underline{T}_2 > 1$. We shall give expressions that allow us to determine a number of epidemiological measures of the infection level in the steady state. We note first that the steady-state expected number of parasites per human host $\underline{X}_1$ is a function of the two transmission factors as follows:

$$X_1(T_1, T_2) = T_1 - \frac{1}{T_2} . \tag{4.1.30}$$

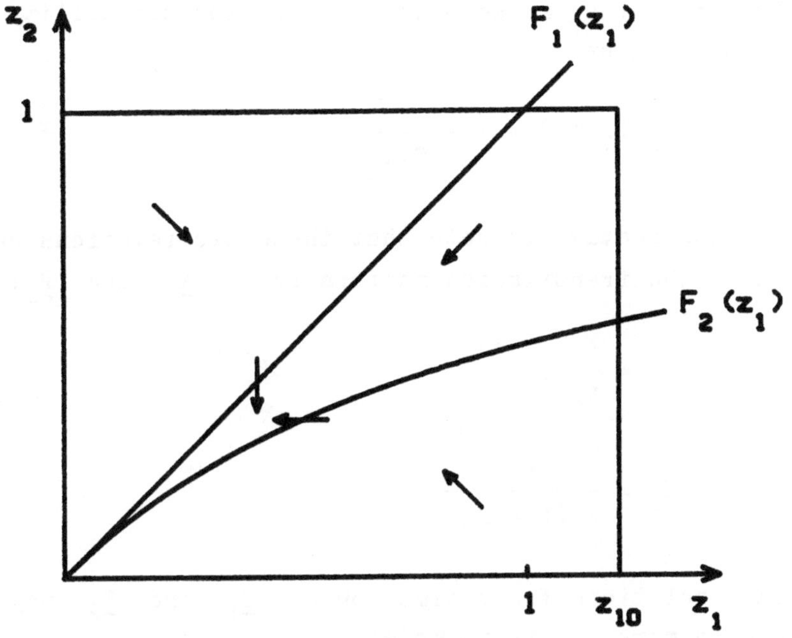

Figure 4.1A. Phase portrait for the model of hermaphroditic helminthiasis below the threshold.

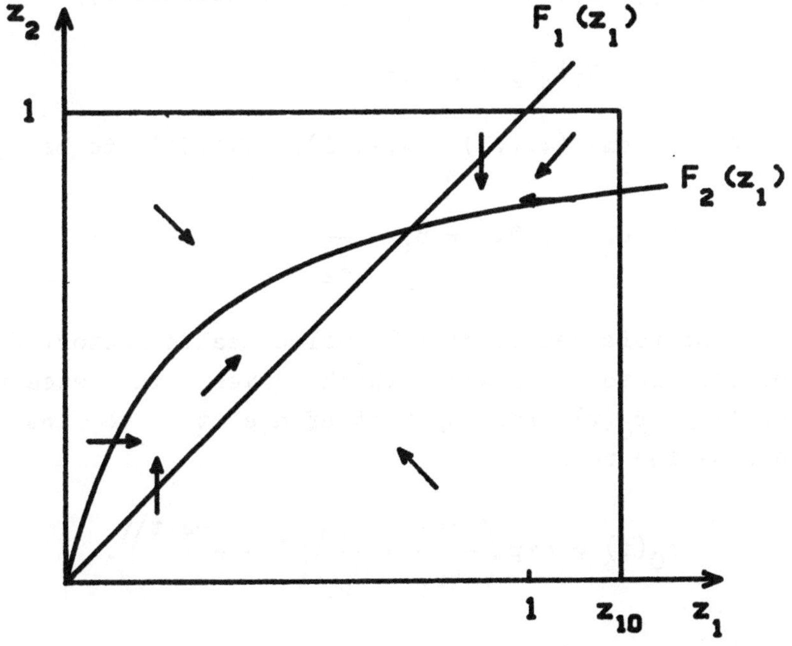

Figure 4.1B. Phase portrait for the model of hermaphroditic helminthiasis above the threshold.

Furthermore, the steady-state infection probability $\underline{P}_2$ for each snail is given by

$$P_2(T_1,T_2) = 1 - \frac{1}{T_1 T_2} . \tag{4.1.31}$$

It is instructive to note that these two relations can be used to determine the transmission factors from $\underline{X}_1$ and $\underline{P}_2$:

$$T_1 = \frac{X_1}{P_2} , \tag{4.1.32}$$

$$T_2 = \frac{P_2}{X_1(1 - P_2)} . \tag{4.1.33}$$

Similar relations for estimation of $\underline{T}_1$ and $\underline{T}_2$ may be expected to hold in more realistic models.

The steady-state human infection rate $\underline{H}_1$ depends on the three parameters $\underline{u}_1$, μ_1, $\underline{T}_2$, where $\underline{u}_1$ is defined by

$$u_1 = \pi_1 \lambda_2 N_2 = \mu_1 T_1 . \tag{4.1.34}$$

It is found from (4.1.3), (4.1.12), (4.1.31) to be equal to

$$H_1(u_1,\mu_1, T_2) = u_1 - \frac{\mu_1}{T_2} . \tag{4.1.35}$$

It is also referred to as the public health factor. We insert this expression into (2.2.43), which gives in steady state the probability $\underline{p}_0(\underline{t})$ for any host of age $\underline{t}$ to be free from infection. The result is

$$p_0(t) = \exp\left(-\left(\frac{u_1}{\mu_1} - \frac{1}{T_2}\right)\left(1 - e^{-\mu_1 t}\right)\right). \tag{4.1.36}$$

Explicit expressions for prevalence, incidence, and recovery probability are found from (2.2.33), (2.2.41), and (2.2.42) to be given by

$$Q(s) = 1 - p_0(s), \qquad (4.1.37)$$

$$I(s,T) = 1 - p_0(T), \qquad (4.1.38)$$

$$R(s,T) = \frac{p_0(s+T) - p_0(s)p_0(T)}{1 - p_0(s)} \quad . \qquad (4.1.39)$$

They are all defined above threshold in the parameter space where $\underline{u}_1, \mu_1, \underline{T}_2$ take their values. In addition, the prevalence $\underline{Q}(\underline{s})$ depends on the age $\underline{s}$, the incidence $I(\underline{s},\underline{T})$ depends on the length $\underline{T}$ of the time interval over which it is defined but is found to be independent of the age $\underline{s}$, while the recovery probability $\underline{R}(\underline{s},\underline{T})$ depends on both $\underline{s}$ and $\underline{T}$.

4.1.5 Control Efficiencies

As in the malaria models there are a number of epidemiological quantities that can be used as indicators of the steady-state health condition in the community. We can identify the parasite load $\underline{X}_1$ (the steady-state expected number of parasites per host), the public health factor $\underline{H}_1$, the prevalence $\underline{Q}(\underline{s})$ at age $\underline{s}$, the incidence $\underline{I}(\underline{s},\underline{T})$ at age $\underline{s}$ over a time interval of length $\underline{T}$, and the recovery probability $\underline{R}(\underline{s},\underline{T})$ at age $\underline{s}$ over a time interval of length $\underline{T}$. Explicit expressions for all of them are given in the previous subsection.

Each of these five quantities can be used to judge the effect of a control action in terms of corresponding control efficiency functions. We confine ourselves here to a study of only two such functions, namely those that measure the efficiency of control of the parasite load $\underline{X}_1$ through reductions of the transmission factors $\underline{T}_1$ and $\underline{T}_2$, respectively. They are denoted by $\underline{C}_1$ and $\underline{C}_2$. Their definitions and explicit expressions are as follows:

$$C_1(T_1,T_2) = \frac{T_1}{X_1}\frac{\partial X_1}{\partial T_1} = \frac{T_1T_2}{T_1T_2-1} \; , \qquad (4.1.40)$$

$$C_2(T_1,T_2) = \frac{T_2}{X_1}\frac{\partial X_1}{\partial T_2} = \frac{1}{T_1T_2-1} \quad . \qquad (4.1.41)$$

It is readily seen that the ratio of the two control efficiency functions is

$$\frac{C_1}{C_2} = T_1 T_2 > 1. \tag{4.1.42}$$

Note also that this ratio can be written in the form

$$\frac{C_1}{C_2} = \frac{1}{1 - P_2} . \tag{4.1.43}$$

This expression for the ratio of the two control efficiency functions is seen to be identical in form to the expression for the corresponding ratio for both the Ross malaria model and the malaria model with superinfection.

We conclude that it is more efficient to reduce the parasite load by reduction of $\underline{T}_1$ than by reduction of $\underline{T}_2$. The advantage is substantial if $\underline{P}_2$ is large, but it is small near the threshold where $\underline{P}_2$ is small.

4.2 A Model for Hermaphroditic Helminthiasis with Human Immunity

A transmission model for hermaphroditic helminthiasis with concomitant immunity of the human beings is readily established by using the superinfection process with immunity of Section 2.3 as host model for the human phase of the life cycle of the parasites. The second model of Section 2.6 serves as host model for the snail phase of the life cycle.

By proceeding as in Section 4.1 we can establish a system of differential equations for the two state variables $\underline{x}_1$ and $\underline{p}_2$, where $\underline{x}_1$ is the expected number of parasites per human host and $\underline{p}_2$ is the snail infection probability. The system of differential equations takes the following form:

$$x_1' = \pi_1\lambda_2N_2p_2(1 - x_1/S) - \mu_1x_1, \tag{4.2.1}$$

$$p_2' = \pi_2\lambda_1N_1x_1(1 - p_2) - \mu_2p_2. \tag{4.2.2}$$

These equations contain nine parameters. Eight of these appear in the model for hermaphroditic helminthiasis without human immunity of Section 4.1, while the ninth one, $\underline{S}$, is the saturation level for each human host.

The next step in the analysis is to introduce suitable transmission factors as functions of the nine basic parameters. We note that the saturation level $\underline{S}$ has the same quasi-dimension as the state variable $\underline{x}_1$. We therefore introduce new state variables free of quasi-dimension by defining

$$z_1 = x_1/S, \tag{4.2.3}$$

$$z_2 = p_2. \tag{4.2.4}$$

It follows then that the state variables $\underline{z}_1$, $\underline{z}_2$ satisfy the following system of differential equations:

$$z_1' = \mu_1(T_1z_2(1-z_1) - z_1), \tag{4.2.5}$$

$$z_2' = \mu_2(T_2z_1(1-z_2) - z_2), \tag{4.2.6}$$

where

$$T_1 = \frac{\pi_1\lambda_2N_2}{\mu_1S} \tag{4.2.7}$$

and

$$T_2 = \frac{\pi_2\lambda_1N_1S}{\mu_2} \quad . \tag{4.2.8}$$

A comparison shows that our equations (4.2.5), (4.2.6) have the same form as equations (3.1.10), (3.1.11) in the Ross model for malaria. We can therefore refer to Section 3.1 for the behaviour of the solutions of our system of equations. In particular, we conclude that the threshold in $\underline{T}_1 - \underline{T}_2$ space is given by the relation $\underline{T}_1\underline{T}_2 = 1$. It follows from the definitions (4.2.7), (4.2.8) of the transmission factors that the threshold condition is independent of the saturation level $\underline{S}$.

The further analysis of this model is straightforward, since we can rely heavily on the results of Section 3.1. For the age-dependence of prevalence, incidence, and recovery probability we use the expressions in Section 2.3.

CHAPTER 5. TRANSMISSION MODELS FOR SCHISTOSOMIASIS

One important difference between schistosomiasis and hermaphroditic helminthiasis is that the sexually mature forms of the schistosomes are dioecious, i.e. male and female forms appear in separate individuals. Mating takes place between males and females, and it is only the mated female parasites that lay eggs. In contrast, all sexually mature parasites lay eggs in hermaphroditic helminthiasis. Thus, the infectivity of the human population in schistosomiasis is measured by the total number of mated female schistosomes harbored by all human hosts.

In this chapter we formulate and study four models for the transmission of schistosomiasis. The need to account for mating between male and female parasites makes the threshold phenomenon more complicated than in the models for hermaphroditic helminthiasis in the previous chapter. The treatment of the basic schistosomiasis model with monogamous mating in Section 5.1 is followed in Section 5.2 by an analysis of a hybrid version of the Macdonald schistosomiasis model. The latter model adds a hypothesis of snail latency to those introduced in the basic model. The model in Section 5.3 allows for snail latency and also for differential mortality and recovery in the snail population. Finally, Section 5.4 deals with a schistosomiasis model with polygamous mating.

5.1 The Basic Schistosomiasis Model

We assume monogamous mating between male and female parasites in each host, as discussed in the host model of Section 2.4, and that mating takes place as soon as partners are available. The model formulation is based on ideas introduced by Macdonald (1965); see Section 5.2 below. A somewhat more rigorous and detailed discussion of the mathematical problems encountered is given by Nåsell and Hirsch (1973).

5.1.1 Model Formulation

The basic schistosomiasis model is similar in structure to the basic model for hermaphroditic helminthiasis developed in the preceding chapter. The eight basic parameters introduced there will serve also for this model; the only difference is that the egg-laying rate λ_1 is in the present model defined as the number of eggs laid per unit time by each mated female parasite.

The model has the form of a collection of Markov chains

$$\{F^{(1)}(t),\ M^{(1)}(t),\ldots,F^{(N_1)}(t),\ M^{(N_1)}(t),\ X_2(t)\}\quad,\quad t \geq 0.$$

To each human host $\underline{k}$ there correspond two Markov chains $\underline{F}^{(k)}$ and $\underline{M}^{(k)}$ that give the number of female and male parasites, respectively, in that host. The state space of each of these Markov chains is the set of all nonnegative integers. The Markov chain $\underline{X}_2$ gives the number of infected snails. Its state space is the set of all nonnegative integers less than or equal to $\underline{N}_2$.

Each $\underline{F}^{(k)}$ and $\underline{M}^{(k)}$ is a superinfection process (see Section 2.2) with infection rate $\frac{1}{2}\,\underline{h}_1(\underline{t})$ and death rate per parasite μ_1. This implies that the total number of parasites in host $\underline{k}$ at time $\underline{t}$, $\underline{X}_1^{(k)}(\underline{t}) = \underline{F}^{(k)}(\underline{t}) + \underline{M}^{(k)}(\underline{t})$, is a superinfection process with infection rate $\underline{h}_1(\underline{t})$ and death rate per parasite μ_1. For $\underline{X}_2$ we use the second host model of Section 2.6 with infection rate $\underline{h}_2(\underline{t})$ and death rate per snail μ_2.

We assume that initially all the random variables $\underline{F}^{(k)}(0)$ and $\underline{M}^{(k)}(0)$ have Poisson distributions with the same parameter $\underline{x}_{10}/2$ and are independent. It follows that the random variables $\underline{X}_1^{(k)}(0)$ also have Poisson distributions with parameter $\underline{x}_{10}$. We assume also that the random variable $\underline{X}_2(0)$ has a binomial distribution with parameters $\underline{N}_2$ and $\underline{p}_{20}$. It follows from these initial assumptions that the number of parasites in human host $\underline{k}$, $\underline{X}^{(k)}(\underline{t})$, has a Poisson distribution for each $\underline{t} > 0$ with time-dependent parameter $\underline{x}_1(\underline{t})$ independent of the individual host, and that the infection probability $\underline{p}_2(\underline{t})$ is the same for all snails. It follows furthermore from (2.2.32) that the expected number of parasites per human host $\underline{x}_1(\underline{t})$ satisfies the differential equation

$$x_1' = h_1(t) - \mu_1 x_1. \qquad (5.1.1)$$

The infection probability $\underline{p}_2$ is found from (2.1.23) to satisfy the differential equation

$$p_2' = h_2(t)(1-p_2) - \mu_2 p_2. \qquad (5.1.2)$$

We proceed to express the infection rates $\underline{h}_1(\underline{t})$ and $\underline{h}_2(\underline{t})$ in terms of state variables and parameters. The infection rate per human being $\underline{h}_1(\underline{t})$ obeys the same expression as in the Section 4.1:

$$h_1(t) = \pi_1 \lambda_2 N_2 p_2(t). \qquad (5.1.3)$$

The infection rate per uninfected snail $\underline{h}_2(\underline{t})$ is equal to $\pi_2 \lambda_1$ multiplied by the total number of mated female parasites in the community. A hybrid assumption is used to replace this stochastic quantity by its expected value. It follows from (2.4.20) that the expected number of mated female parasites at time $\underline{t}$ in each host is equal to $\frac{1}{2}\Psi(\underline{x}_1(\underline{t}))$, where the function Ψ is defined by (2.4.21). Hence the infection rate $\underline{h}_2(\underline{t})$ can be written

$$h_2(t) = \frac{1}{2}\pi_2 \lambda_1 N_1 \Psi(x_1(t)). \qquad (5.1.4)$$

By inserting these expressions for the infection rates into equations (5.1.1), (5.1.2), we are led to the following equations for the state variables $\underline{x}_1$, $\underline{p}_2$:

$$x_1' = \pi_1 \lambda_2 N_2 p_2 - \mu_1 x_1, \qquad (5.1.5)$$

$$p_2' = \frac{1}{2}\pi_2 \lambda_1 N_1 \Psi(x_1)(1-p_2) - \mu_2 p_2. \qquad (5.1.6)$$

The initial values $\underline{x}_{10}$, $\underline{p}_{20}$ obey the same constraints as in Section 4.1, namely $0 \le \underline{x}_{10}$, $0 \le \underline{p}_{20} \le 1$.

The quasi-dimensional analysis carried out in Subsection 4.1.2 applies to the case treated here. The only additional observation required is that the quasi-dimension of the expected number of mated parasites per host, $\Psi(\underline{x}_1)$, is the same as the quasi-dimension of the expected number of parasites per host $\underline{x}_1$, and thus equal to $\underline{P}\ \underline{H}^{-1}$. Based on the results in Subsection 4.1.2 we introduce two transmission factors $\underline{T}_1$ and $\underline{T}_2$ by writing

$$T_1 = \frac{\pi_1 \lambda_2 N_2}{\mu_1} \tag{5.1.7}$$

and

$$T_2 = \frac{\pi_2 \lambda_1 N_1}{\mu_2} \ . \tag{5.1.8}$$

Furthermore, we put

$$z_1 = x_1 / T_1 \tag{5.1.9}$$

and

$$z_2 = p_2. \tag{5.1.10}$$

Thus we are led to the following system of differential equations for the state variables $\underline{z}_1$ and $\underline{z}_2$:

$$z_1' = \mu_1(z_2 - z_1), \tag{5.1.11}$$

$$z_2' = \mu_2(\tfrac{1}{2}\, T_2 \Psi(T_1 z_1)(1-z_2) - z_2) \ . \tag{5.1.12}$$

Initial values for the two functions $\underline{z}_1$ and $\underline{z}_2$ are given at $\underline{t} = 0$ by $\underline{z}_{10} = \underline{x}_{10}/\underline{T}_1$ and $\underline{z}_{20} = \underline{p}_{20}$. The initial point belongs to the rectangle $\underline{R}$ defined by

$$R = \{(z_1, z_2) : 0 \leq z_1 \leq \hat{z}_1,\ \ 0 \leq z_2 \leq 1\} \ , \tag{5.1.13}$$

where

$$\hat{z}_1 = \max(1, z_{10}). \tag{5.1.14}$$

We show in Subsection 5.1.4 below that the rectangle $\underline{R}$ is positively invariant.

5.1.2 Critical Points and Threshold Functions

In this subsection we shall analyze the critical points of the system of equations (5.1.11), (5.1.12) and determine a threshold function in the form of a relation between the transmission factors T_1 and T_2. The isocline where $z_1' = 0$ is the straight line

$$z_2 = F_1(z_1) = z_1, \tag{5.1.15}$$

while the isocline where $z_2' = 0$ is given by

$$z_2 = F_2(z_1) = \frac{T_2\Psi(T_1 z_1)}{2 + T_2\Psi(T_1 z_1)} . \tag{5.1.16}$$

Critical points are found as intersections of the two isoclines. By denoting any critical point by $(\bar{z}_1, \bar{z}_2)$, we find from the equations for the isoclines that

$$T_2\Psi(T_1\bar{z}_1)(1 - \bar{z}_1) = 2\bar{z}_1 \tag{5.1.17}$$

and that

$$\bar{z}_2 = \bar{z}_1 . \tag{5.1.18}$$

We shall use relation (5.1.17) to determine $\bar{z}_1$ as a function of the transmission factors T_1 and T_2. It turns out that the analysis of this transcendental equation is somewhat simplified if we introduce $\bar{x}_1 = T_1\bar{z}_1$ to denote the steady-state expected number of parasites per host. Equation (5.1.17) then takes the form

$$T_2\Psi(\bar{x}_1)(T_1 - \bar{x}_1) = 2\bar{x}_1 . \tag{5.1.19}$$

We note first that $\Psi(0) = 0$, and that therefore $\bar{x}_1 = 0$ is a solution of (5.1.19) for any positive values of the transmission factors T_1 and T_2. Thus, the origin is always a critical point of the system of equations (5.1.11), (5.1.12). We note also that $\Psi(x) \geq 0$ for any real x and that therefore any solution $\bar{x}_1$ of (5.1.19) must satisfy $0 \leq \bar{x}_1 < T_1$. We shall show that there exists a threshold function $\bar{\bar{T}}_1$ with the property that if $T_1 < \bar{\bar{T}}_1(T_2)$, then (5.1.19) has only one solution, namely $x_1 = 0$. On the other hand, if $T_1 > \bar{\bar{T}}_1(T_2)$, then

(5.1.19) has two additional solutions. These two additional solutions coincide on the threshold where $\underline{T}_1 = \overset{=}{\underline{T}}_1(\underline{T}_2)$.

For each fixed value of $\underline{T}_2 > 0$, we define the function $\hat{\underline{T}}_1(.;\underline{T}_2)$ by setting

$$\hat{T}_1(x;T_2) = x + \frac{2x}{T_2\Psi(x)}, \quad 0 < x. \tag{5.1.20}$$

Furthermore, for each fixed value of $\underline{T}_1 > 0$, we define the function $\hat{T}_2(\ .\ ;\ T_1)$ by

$$\hat{T}_2(x\ ;\ T_1) = \frac{2x}{(T_1-x)\Psi(x)}, \quad 0 < x < T_1. \tag{5.1.21}$$

We shall show that these functions have the behaviour shown in Figures 5.1A and 5.1B. Specifically, this means that $\hat{\underline{T}}_1(\underline{x};\underline{T}_2)$, viewed as a function of $\underline{x}$ for fixed value of $\underline{T}_2$, has a minimum at an $\underline{x}$-value that clearly depends on $\underline{T}_2$. The $\underline{x}$-value at which the minimum occurs is denoted $\bar{\underline{X}}_2(\underline{T}_2)$. Furthermore, the minimum value also depends on $\underline{T}_2$ and is denoted $\overset{=}{\underline{T}}_1(\underline{T}_2)$. Similarly, the function $\hat{\underline{T}}_2(.;\underline{T}_1)$ has a minimum at $\underline{x} = \bar{\underline{X}}_1(\underline{T}_1)$ and the minimum value is denoted $\overset{=}{\underline{T}}_2(\underline{T}_1)$.

These properties of the functions $\hat{\underline{T}}_1$ and $\hat{\underline{T}}_2$ are equivalent to a description of the threshold phenomenon. Indeed, for any $\bar{\underline{x}}_1 > 0$ we find that equation (5.1.19) is equivalent to

$$T_1 = \hat{T}_1(\bar{x}_1;\ T_2) \tag{5.1.22}$$

and to

$$T_2 = \hat{T}_2(\bar{x}_1;\ T_1). \tag{5.1.23}$$

For given positive values of $\underline{T}_1$, $\underline{T}_2$, there is no solution of (5.1.22) unless $\underline{T}_1$ is larger than or equal to the minimum value of $\hat{\underline{T}}_1(\underline{x};\underline{T}_2)$, i.e. unless $\underline{T}_1 \geq \overset{=}{\underline{T}}_1(\underline{T}_2)$. This in turn means that there is no critical point of the system (5.1.11), (5.1.12) other than the

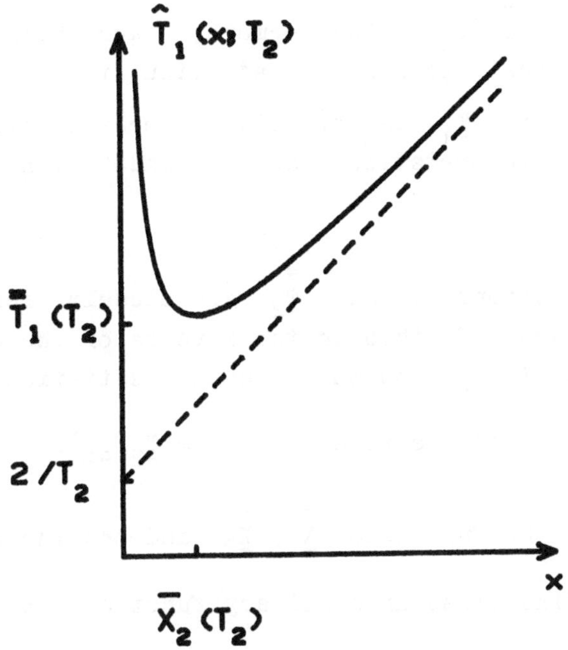

Figure 5.1A. The function $\hat{\underline{T}}_1(\underline{x};\underline{T}_2)$ is plotted as a function of $\underline{x}$ for fixed value of $\underline{T}_2$.

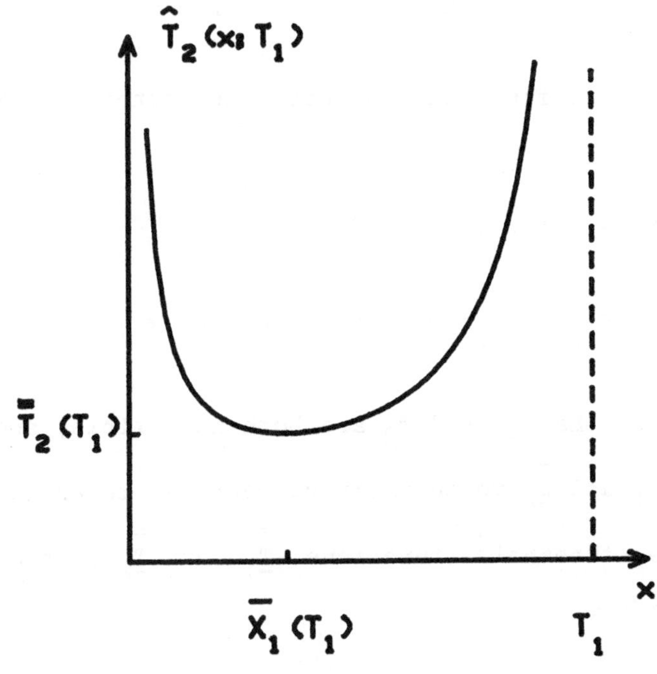

Figure 5.1B. The function $\hat{\underline{T}}_2(\underline{x};\underline{T}_1)$ is plotted as a function of $\underline{x}$ for fixed value of $\underline{T}_1$.

origin unless $\underline{T}_1 > \underline{\underline{T}}_1(\underline{T}_2)$. Furthermore, Figure 5.1A shows that there are two critical points in addition to the origin if $\underline{T}_1 > \underline{\underline{T}}_1(\underline{T}_2)$, i.e. if $\underline{T}_1$ is strictly larger than its threshold value. A similar interpretation can be made with regard to solutions of (5.1.23).

The two threshold functions $\underline{\underline{T}}_1$ and $\underline{\underline{T}}_2$ are closely related; we shall show that each of them is the inverse of the other. This means that any point $(\underline{T}_1, \underline{T}_2)$ on the threshold satisfies both the relation $\underline{T}_1 = \underline{\underline{T}}_1(\underline{T}_2)$ and the relation $\underline{T}_2 = \underline{\underline{T}}_2(\underline{T}_1)$.

We proceed to define the functions $\bar{\underline{X}}_1$, $\bar{\underline{X}}_2$ and the threshold functions $\underline{\underline{T}}_1$, $\underline{\underline{T}}_2$. We define first an auxiliary function $\underline{h}$ by putting

$$h(x) = \begin{cases} \dfrac{e^x - I_0(x)}{I_1(x)}, & x > 0, \\ 2, & x = 0. \end{cases} \tag{5.1.24}$$

With the aid of this function, we define two functions $\bar{\underline{T}}_1$ and $\bar{\underline{T}}_2$ as follows:

$$\bar{T}_1(x) = x\, h(x), \qquad x \geq 0, \tag{5.1.25}$$

$$\bar{T}_2(x) = \frac{2}{(h(x) - 1)\Psi(x)}, \qquad x > 0. \tag{5.1.26}$$

We shall prove that $\bar{\underline{T}}_1$ and $\bar{\underline{T}}_2$ are both invertible, and we define the functions $\bar{\underline{X}}_1$ and $\bar{\underline{X}}_2$ to be their respective inverses. Finally, we define the threshold functions $\underline{\underline{T}}_1$ and $\underline{\underline{T}}_2$ by composition as follows:

$$\bar{\bar{T}}_1 = \bar{T}_1 \circ \bar{X}_2, \tag{5.1.27}$$

$$\bar{\bar{T}}_2 = \bar{T}_2 \circ \bar{X}_1. \tag{5.1.28}$$

Each of them is defined on the domain of all positive real numbers. We shall prove that they are both invertible. It follows then from the defining relations that each is the inverse of the other. Indeed, the inverse of $\bar{\bar{T}}_1$ can be written

$$\bar{\bar{T}}_1^{-1} = \bar{X}_2^{-1} \circ \bar{T}_1^{-1} = \bar{T}_2 \circ \bar{X}_1 = \bar{\bar{T}}_2. \tag{5.1.29}$$

We now embark on the program of first studying the properties of the functions $\bar{\underline{X}}_1$, $\bar{\underline{X}}_2$, $\bar{\bar{\underline{T}}}_1$, $\bar{\bar{\underline{T}}}_2$ defined above and then showing that they actually describe the behaviour of the function $\hat{\underline{T}}_1$ and $\hat{\underline{T}}_2$ as indicated in Figures 5.1A and 5.1B.

We begin by developing some properties of the auxiliary function $\underline{h}$. We note from the power series expression for the modified Bessel function $\underline{I}_k(\underline{x})$ in (2.4.6) that $\underline{I}_k(\underline{x}) > 0$ for $\underline{x} > 0$ and $\underline{k}$ = 0,1,2,... . Hence we get from (2.4.8) that

$$e^x - I_0(x) > 2I_1(x), \quad x > 0. \tag{5.1.30}$$

It follows from the definition of $\underline{h}$ in (5.1.24) that

$$h(x) > 2, \quad x > 0. \tag{5.1.31}$$

By using (2.4.6) and the power series expansion of the exponential function we find that

$$h(x) \sim 2, \quad x \to 0. \tag{5.1.32}$$

This result shows that $\underline{h}$ is a continuous function. For large values of $\underline{x}$ we use the asymptotic expression for the modified Bessel function $\underline{I}_k(\underline{x})$ in (2.4.11) to conclude that

$$h(x) \sim (2\pi x)^{1/2}, \quad x \to \infty . \tag{5.1.33}$$

Finally, we prove that $\underline{h}$ is a monotonically increasing function. By differentiation of the defining relation (5.1.24) we find that

$$h'(x) = \frac{g(x)}{I_1^2(x)}, \tag{5.1.34}$$

where

$$g(x) = (e^x - I_1(x))I_1(x) - (e^x - I_0(x))\tfrac{1}{2}(I_0(x)+I_2(x)). \tag{5.1.35}$$

Since $I_0(0) = 1$ and $I_1(0) = 0$, we conclude that $g(0) = 0$. In order to prove that $h'(x) > 0$ for $x > 0$ it suffices to show that $g'(x) > 0$ for $x > 0$. After differentiation and simplification using (2.4.7), (2.4.9), (2.4.10) we find that

$$g'(x) = \frac{1}{x}\,(e^x - I_0(x))\, I_2(x) + \frac{1}{x}\, I_1^2(x) > 0, \qquad x > 0. \tag{5.1.36}$$

This proves that

$$h'(x) > 0, \quad x > 0. \tag{5.1.37}$$

We consider now the function $\bar{T}_1 : [0, \infty) \to [0, \infty)$ defined by (5.1.25). We find by differentiation and application of the inequalities for h and h' in (5.1.31), (5.1.37) that

$$\bar{T}_1'(x) = h(x) + x\, h'(x) > 2, \qquad x > 0. \tag{5.1.38}$$

Hence $\bar{T}_1$ is injective. To prove that $\bar{T}_1$ is surjective, note that

$$\bar{T}_1(x) \sim 2x, \qquad x \to 0 \tag{5.1.39}$$

and that

$$\bar{T}_1(x) \sim (2\pi x^3)^{1/2}, \quad x \to \infty\ . \tag{5.1.40}$$

Here we have used the asymptotic expressions for h in (5.1.32) and (5.1.33). We conclude that $\bar{T}_1$ maps onto $[0, \infty)$. Hence $\bar{T}_1$ is bijective and its inverse $\bar{X}_1 : [0, \infty) \to [0, \infty)$ exists. From the lower bound for the derivative of $\bar{T}_1$ in (5.1.38) we get

$$0 < \bar{X}_1'(T_1) < 1/2, \quad T_1 > 0. \tag{5.1.41}$$

From the mean value theorem we conclude that

$$0 < \bar{X}_1(T_1) < T_1/2, \quad T_1 > 0. \tag{5.1.42}$$

The asymptotic expressions for the function $\bar{T}_1$ in (5.1.39), (5.1.40) show that the inverse function $\bar{X}_1$ has the following asymptotic behaviour:

$$\bar{X}_1(T_1) \sim T_1/2, \quad T_1 \to 0, \tag{5.1.43}$$

$$\bar{X}_1(T_1) \sim (T_1^2/(2\pi))^{1/3}, \quad T_1 \to \infty . \tag{5.1.44}$$

We consider next the function $\bar{T}_2 : (0, \infty) \to (0, \infty)$ defined by (5.1.26). By differentiation and application of the inequalities for h and h' in (5.1.31), (5.1.37) and the lower bounds for Ψ and Ψ' in (2.4.24), (2.4.25) we find that

$$\bar{T}_2'(x) = -2 \frac{h'(x)\Psi(x)+(h(x)-1)\Psi'(x)}{((h(x)-1)\Psi(x))^2} < 0, \ x>0. \tag{5.1.45}$$

Hence $\bar{T}_2$ is injective. To prove that $\bar{T}_2$ is also surjective, we consider its asymptotic behaviour for small and large values of x. By using the asymptotic expressions for Ψ in (2.4.29), (2.4.30) and those for h in (5.1.32), (5.1.33) we find that

$$\bar{T}_2(x) \sim 4/x^2, \quad x \to 0, \tag{5.1.46}$$

$$\bar{T}_2(x) \sim (2/(\pi x^3))^{1/2}, \quad x \to \infty . \tag{5.1.47}$$

We conclude that $\bar{T}_2$ maps onto $(0, \infty)$ and that therefore $\bar{T}_2$ is bijective. Hence its inverse $\bar{X}_2$: $(0, \infty) \to (0, \infty)$ exists. The asymptotic expressions for $\bar{X}_2$ follow from those for $\bar{T}_2$ in (5.1.46), (5.1.47). The results are

$$\bar{X}_2(T_2) \sim (2/(\pi T_2^2))^{1/3}, \quad T_2 \to 0, \tag{5.1.48}$$

$$\bar{X}_2(T_2) \sim 2\, T_2^{-1/2}, \quad T_2 \to \infty. \tag{5.1.49}$$

Furthermore,

$$\bar{X}_2'(T_2) = \frac{1}{\bar{T}_2'(\bar{X}_2(T_2))} < 0, \quad T_2 > 0. \tag{5.1.50}$$

We consider now the threshold functions $\bar{\bar{T}}_1$ and $\bar{\bar{T}}_2$ defined by (5.1.27), (5.1.28). For $T_2 > 0$ we find that

$$\bar{\bar{T}}_1'(T_2) = \bar{T}_1'(\bar{X}_2(T_2))\bar{X}_2'(T_2) < 0\ , \tag{5.1.51}$$

where the inequality follows from an application of (5.1.38), (5.1.50). The asymptotic behaviour of $\bar{\bar{T}}_1$ for small and large values of its argument follows from the asymptotic behaviour of $\bar{T}_1$ in (5.1.39), (5.1.40) and of $\bar{X}_2$ in (5.1.48), (5.1.49). The results are

$$\bar{\bar{T}}_1(T_2) \sim 2/T_2, \quad T_2 \to 0\ , \tag{5.1.52}$$

$$\bar{\bar{T}}_1(T_2) \sim 4T_2^{-1/2}, \quad T_2 \to \infty\ . \tag{5.1.53}$$

The inequality in (5.1.51) shows that $\bar{\bar{T}}_1$ is injective and the asymptotic expressions above show that $\bar{\bar{T}}_1$ is surjective. It follows that its inverse exists. We have already shown that $\bar{\bar{T}}_2$ is then the inverse of $\bar{\bar{T}}_1$. It follows that

$$\bar{\bar{T}}_2'(T_1) = \frac{1}{\bar{\bar{T}}_1'(\bar{\bar{T}}_2(T_1))} < 0, \quad T_1 > 0, \tag{5.1.54}$$

$$\bar{\bar{T}}_2(T_1) \sim 16/T_1^2, \quad T_1 \to 0, \tag{5.1.55}$$

$$\bar{\bar{T}}_2(T_1) \sim 2/T_1, \quad T_1 \to \infty. \tag{5.1.56}$$

The threshold function is a boundary in parameter space between that domain (below the threshold) where the equations (5.1.11),(5.1.12) have the origin as the only critical points and that domain (above the threshold) where there are two additional critical points.

This concludes the derivation of the basic properties of the four functions $\bar{\underline{X}}_1$, $\bar{\underline{X}}_2$, $\bar{\bar{\underline{T}}}_1$, $\bar{\bar{\underline{T}}}_2$. We proceed to show that these functions describe the behaviour of the functions $\hat{\underline{T}}_1$ and $\hat{\underline{T}}_2$ as shown in Figures 5.1A and 5.1B.

We note first from the definition of the function Ψ in (2.4.21) and the expression for its derivative in (2.4.22) that

$$x\Psi'(x) - \Psi(x) = x\, e^{-x} I_1(x) > 0, \quad x > 0 . \qquad (5.1.57)$$

It follows that the function $\underline{h}$ can be expressed in terms of Ψ and Ψ' in the following two ways:

$$h(x) = 1 + \frac{\Psi(x)}{x\Psi'(x) - \Psi(x)}, \quad x > 0, \qquad (5.1.58)$$

$$h(x) = \frac{x\Psi'(x)}{x\Psi'(x) - \Psi(x)}, \quad x > 0. \qquad (5.1.59)$$

By differentiation of (5.1.20) and application of (5.1.58) and the defining relation (5.1.26) for the function $\bar{\underline{T}}_2$ we find that

$$\frac{\partial \hat{T}_1(x;T_2)}{\partial x} = 1 - \frac{\bar{T}_2(x)}{T_2}, \quad x > 0. \qquad (5.1.60)$$

In a similar way we find by differentiation of (5.1.21) and application of (5.1.59), (5.1.25), (5.1.26) that

$$\frac{\partial \hat{T}_2(x;T_1)}{\partial x} = \frac{\bar{T}_2(x)}{(T_1-x)^2}\,(\bar{T}_1(x) - T_1), \quad 0 < x < T_1. \qquad (5.1.61)$$

Relation (5.1.60) shows that the partial derivative of $\hat{\underline{T}}_1(\underline{x};\underline{T}_2)$ with respect to $\underline{x}$ is equal to zero for an $\underline{x}$-value that satisfies $\bar{\underline{T}}_2(\underline{x}) = \underline{T}_2$, i.e. for

$$x = \bar{X}_2(T_2) \ . \qquad (5.1.62)$$

We have shown in (5.1.45) that $\bar{T}_2$ is a monotonically decreasing function of $\underline{x}$. It follows therefore from (5.1.60) that $\hat{\underline{T}}_2$ has a minimum at the $\underline{x}$-value given in (5.1.62). Furthermore, the minimum value is equal to

$$\hat{T}_1(\bar{X}_2(T_2);T_2) = \bar{X}_2(T_2)\ (1 + \frac{2}{T_2\Psi(\bar{X}_2(T_2))}). \qquad (5.1.63)$$

This expression can be further simplified. By putting $\underline{x} = \bar{\underline{X}}_2(\underline{T}_2)$ in (5.1.26) we get

$$h(\bar{X}_2(T_2)) = 1 + \frac{2}{T_2\Psi(\bar{X}_2(T_2))} \ . \qquad (5.1.64)$$

Thus we get, using (5.1.25) and (5.1.27), that the minimum value of $\hat{\underline{T}}_1$ is equal to the threshold value determined by $\underline{T}_2$:

$$\hat{T}_1(\bar{X}_2(T_2);T_2) = \bar{\bar{T}}_1(T_2), \qquad (5.1.65)$$

as indicated in Figure 5.1A.

In a similar way we find from relation (5.1.61) that the derivative of $\hat{\underline{T}}_2(\ .\ ;\ \underline{T}_1)$ is equal to zero for an $\underline{x}$-value that satisfies $\bar{\underline{T}}_1(\underline{x}) = \underline{T}_1$, i.e. for

$$x = \bar{X}_1(T_1). \qquad (5.1.66)$$

We have shown in (5.1.38) that $\bar{\underline{T}}_1$ is a monotonically increasing function. It follows therefore from (5.1.61) that $\hat{\underline{T}}_2$ has a minimum at the $\underline{x}$-value given by (5.1.66). The minimum value is equal to

$$\hat{T}_2(\bar{X}_1(T_1);T_1) = \frac{2\bar{X}_1(T_1)}{(T_1-\bar{X}_1(T_1))\Psi(\bar{X}_1(T_1))} \quad . \tag{5.1.67}$$

We simplify this expression. By putting $\underline{x} = \bar{\underline{X}}_1(\underline{T}_1)$ in (5.1.25) we find that

$$T_1 = \bar{X}_1(T_1)h(\bar{X}_1(T_1)). \tag{5.1.68}$$

Thus we find, using (5.1.26) and (5.1.28), that the minimum value of $\hat{\underline{T}}_2(\underline{x};\underline{T}_1)$ is equal to

$$\hat{T}_2(\bar{X}_1(T_1);T_1) = \bar{\bar{T}}_2(T_1). \tag{5.1.69}$$

The proof of the qualitative features of the function $\hat{\underline{T}}_1$ and $\hat{\underline{T}}_2$ indicated in Figures 5.1A and B has now been completed, and the threshold function has been defined and some of its properties have been given. We proceed to study the critical points of the system of equations (5.1.11), (5.1.12) and to show how their coordinates are determined as functions of the transmission factors $\underline{T}_1$ and $\underline{T}_2$.

Below the threshold, i.e. for $\underline{T}_1 < \bar{\bar{\underline{T}}}_1(\underline{T}_2)$, or, equivalently, $\underline{T}_2 < \bar{\bar{\underline{T}}}_2(\underline{T}_1)$, the origin is the only critical point. Above the threshold there are two additional critical points for which $\bar{\underline{x}}_1$ satisfies (5.1.22) and (5.1.23). The function $\hat{\underline{T}}_1(\,.\, ; \underline{T}_2)$ is monotonically decreasing on the interval $(0, \bar{X}_2(T_2)]$ and monotonically increasing on the interval $[\bar{\underline{X}}_2(\underline{T}_2), \infty)$. The function is therefore not invertible but it is easy to see that its two restrictions to these two intervals are. We denote the two restrictions by $\hat{\underline{T}}_{11}(\,.\, ; \underline{T}_2)$ and $\hat{\underline{T}}_{12}(\,.\, ; \underline{T}_2)$, respectively. The domain and range of these two functions are

$$\hat{T}_{11}(\,.\, ; T_2) : (0, \bar{X}_2(T_2)] \to [\bar{\bar{T}}_1(T_2), \infty) \tag{5.1.70}$$

and

$$\hat{T}_{12}(\ , \ ; T_2) : [\bar{X}_2(T_2), \infty) \to [\bar{\bar{T}}_1(T_2), \infty). \qquad (5.1.71)$$

The value of either of these functions is determined by the expression on the right-hand side of (5.1.20). The first of these functions is monotonically decreasing and the second one is monotonically increasing. Therefore, both of them are injective. Furthermore, both of them take the value $\bar{\bar{\underline{T}}}_1(\underline{T}_2)$ for $\underline{x} = \bar{\underline{X}}_2(\underline{T}_2)$. By using the asymptotic expressions for Ψ in (2.4.29), (2.4.30) we find that

$$\hat{T}_{11}(x;T_2) \sim 4/(T_2 x), \quad x \to 0 \qquad (5.1.72)$$

and

$$\hat{T}_{12}(x;T_2) \sim x, \qquad x \to \infty. \qquad (5.1.73)$$

We conclude that both of the functions are surjective and hence bijective and invertible. Their inverses are denoted by $\underline{X}_1(\ . \ ,T_2)$ and $\underline{X}_2(\ . \ , \underline{T}_2)$, respectively. The domain and range of these functions are:

$$X_1(\ . \ , T_2) : [\bar{\bar{T}}_1(T_2), \infty) \to (0,\bar{X}_2(T_2)] \qquad (5.1.74)$$

and

$$X_2(\ . \ , T_2) : [\bar{\bar{T}}_1(T_2), \infty) \to [\bar{X}_2(T_2), \infty). \qquad (5.1.75)$$

The first of these functions is monotonically decreasing and the second is monotonically increasing.

In the following we shall consider the functions $\underline{X}_1$ and $\underline{X}_2$ defined on the domain

$$\{(T_1,T_2) : T_2 > 0, \quad T_1 \geq \bar{\bar{T}}_1(T_2)\}. \qquad (5.1.76)$$

Similar arguments can be used to define two restrictions of the function $\hat{\underline{T}}_2 (\ . \ ; \underline{T}_1)$ as follows:

$$\hat{T}_{21}(\ . \ ; T_1) : (0,\bar{X}_1(T_1)] \to [\bar{\bar{T}}_2(T_1), \infty), \qquad (5.1.77)$$

$$\hat{T}_{22}(\,\cdot\,;T_1) : [\bar{X}_1(T_1),\ T_1) \rightarrow [\bar{\bar{T}}_2(T_1),\ \infty). \qquad (5.1.78)$$

These two functions are invertible. Their inverses are denoted by $\tilde{\underline{X}}_1(\underline{T}_1,\,\cdot\,)$ and $\tilde{\underline{X}}_2(\underline{T}_1,\,\cdot\,)$, respectively. Their domain and range are

$$\tilde{X}_1(T_1,\,\cdot\,) : [\bar{\bar{T}}_2(T_1),\ \infty) \rightarrow (0,\bar{X}_1(T_1)]\ , \qquad (5.1.79)$$

$$\tilde{X}_2(T_1,\,\cdot\,) : [\bar{\bar{T}}_2(T_1),\ \infty) \rightarrow [\bar{X}_1(T_1),T_1). \qquad (5.1.80)$$

The first one is monotonically decreasing and the second one is monotonically increasing. The functions $\tilde{\underline{X}}_1$ and $\tilde{\underline{X}}_2$ are both defined on the domain

$$\{(T_1,T_2) : T_1 > 0,\quad T_2 \geq \bar{\bar{T}}_2(T_1)\}. \qquad (5.1.81)$$

We proceed to prove that $\tilde{\underline{X}}_1 = \underline{X}_1$ and that $\tilde{\underline{X}}_2 = \underline{X}_2$. First we note that the domains of definition of these four functions are all the same. Indeed, since $\bar{\bar{\underline{T}}}_2$ is a decreasing function and since $\bar{\bar{\underline{T}}}_1$ and $\bar{\bar{\underline{T}}}_2$ are inverses of each other, we find that $\underline{T}_1 \geq \bar{\bar{\underline{T}}}_1(\underline{T}_2)$ implies that $\bar{\bar{\underline{T}}}_2(\underline{T}_1) \leq \bar{\bar{\underline{T}}}_2(\bar{\bar{\underline{T}}}_1(\underline{T}_2)) = \underline{T}_2$. Similarly, $\underline{T}_2 \geq \bar{\bar{\underline{T}}}_2(\bar{\bar{\underline{T}}}_1)$ implies that $\bar{\bar{\underline{T}}}_1(\underline{T}_2) \leq \underline{T}_1$. Next we note that for any point $(\underline{T}_1,\underline{T}_2)$ on or above the threshold we have $\bar{\underline{X}}_2(\underline{T}_2) \leq \bar{\underline{X}}_1(\underline{T}_1)$. Since $\bar{\underline{X}}_2$ is a decreasing function we find that $\underline{T}_2 \geq \bar{\bar{\underline{T}}}_2(\underline{T}_1)$ implies $\bar{\underline{X}}_2(\underline{T}_2) \leq \bar{\underline{X}}_2(\bar{\bar{\underline{T}}}_2(\underline{T}_1)) = \bar{\underline{X}}_2(\bar{\underline{T}}_2(\bar{\underline{X}}_1(\underline{T}_1))) = \bar{\underline{X}}_1(\underline{T}_1)$.

Now, $\underline{X}_1(\underline{T}_1,\underline{T}_2)$ is the unique solution of (5.1.22) that satisfies $0 < \underline{X}_1(\underline{T}_1,\underline{T}_2) < \bar{\underline{X}}_2(\underline{T}_2)$, while $\tilde{\underline{X}}_1(\underline{T}_1,\underline{T}_2)$ is the unique solution of (5.1.23) that satisfies $0 < \tilde{\underline{X}}_1(\underline{T}_1,\underline{T}_2) < \bar{\underline{X}}_1(\underline{T}_1)$. The two equations are identical and the second existence interval contains the first one. Thus we conclude that $\tilde{\underline{X}}_1 = \underline{X}_1$. The proof that $\tilde{\underline{X}}_2 = \underline{X}_2$ is similar. The tilde in the notation for the functions $\tilde{\underline{X}}_1$ and $\tilde{\underline{X}}_2$ will henceforth be omitted.

This completes the investigation of the critical points of the system of differential equations (5.1.11), (5.1.12). The results are summarized in Figures 5.1C and 5.1D, which show the possible values of the steady-state expected number of parasites per human being as functions of the transmission factors $\underline{T}_1$ and $\underline{T}_2$.

5.1.3 Local Stability of the Critical Points

The system of differential equations (5.1.11), (5.1.12) is for brevity written $\underline{z}' = \underline{f}(\underline{z})$ in vector form. The local stability of any critical point is determined by the signs of the real parts of the eigenvalues of the Jacobian matrix evaluated at the critical point, as discussed in Appendix IV. The Jacobian matrix at an arbitrary point $\underline{z}$ is equal to

$$\frac{\partial f}{\partial z}(z) = \begin{pmatrix} -\mu_1 & \mu_1 \\ \frac{1}{2}\mu_2 T_1 T_2 (1-z_2)\Psi'(T_1 z_1) & -\mu_2(1+\frac{1}{2}T_2\Psi(T_1 z_1)) \end{pmatrix}. \quad (5.1.82)$$

At the origin we find that

$$\frac{\partial f}{\partial z}(0) = \begin{pmatrix} -\mu_1 & \mu_1 \\ 0 & -\mu_2 \end{pmatrix}. \quad (5.1.83)$$

This matrix has the two eigenvalues $-\mu_1$ and $-\mu_2$. Since both of them are negative, we conclude that the origin is asymptotically stable for all parameter values.

Above the threshold there exist two additional critical points that coincide on the threshold. We use the term worm load to refer to the expected number of parasites per human host. The steady-state worm loads at the two additional critical points are equal to $\underline{X}_1(\underline{T}_1,\underline{T}_2)$ and $\underline{X}_2(\underline{T}_1,\underline{T}_2)$, respectively. We use the notation $\underline{\bar{z}}^i$ $\underline{i} = 1,2$, to denote the corresponding critical points of the system of equations (5.1.11), (5.1.12). Thus, we have

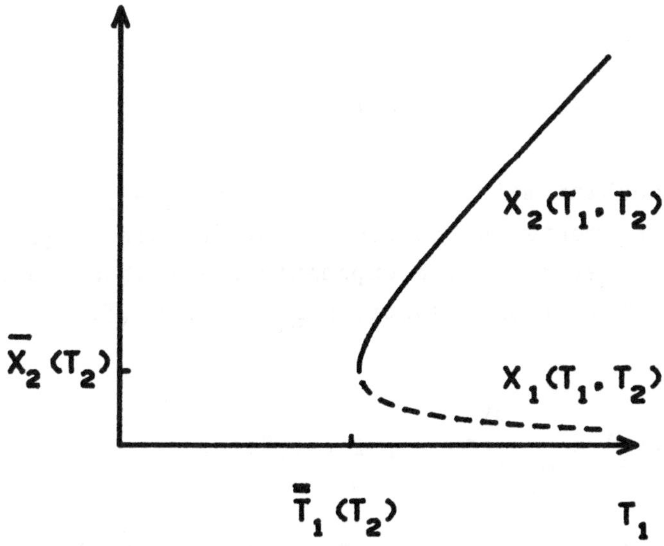

Figure 5.1C. The steady-state expected number of parasites per human being are shown as functions of T_1 for fixed value of T_2.

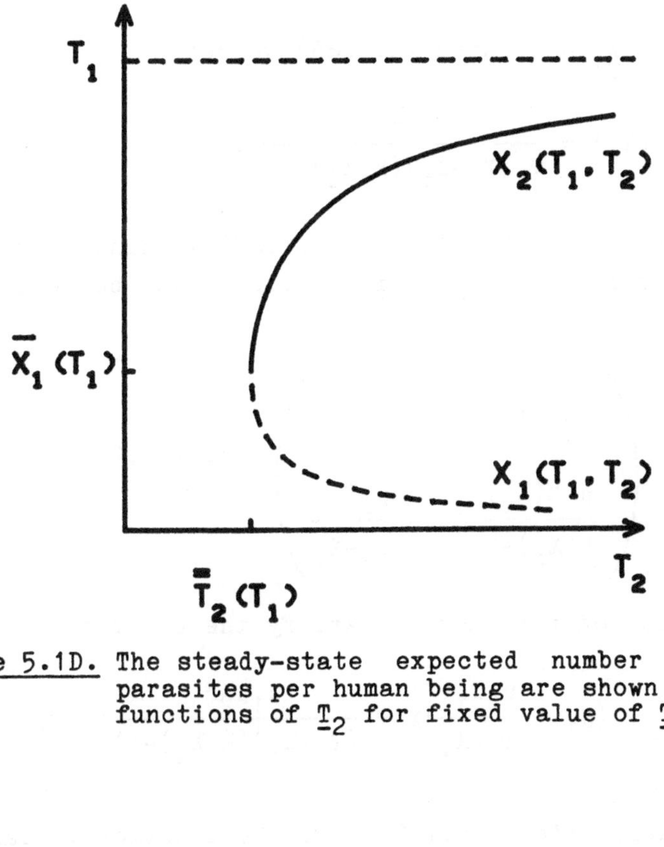

Figure 5.1D. The steady-state expected number of parasites per human being are shown as functions of T_2 for fixed value of T_1.

$$\bar{z}^i = (X_i/T_1,\ X_i/T_1), \quad i = 1,2, \tag{5.1.84}$$

where the arguments of the functions X_i are deleted for brevity.

When the Jacobian matrix is evaluated at $\underline{z} = \bar{z}^i$, $\underline{i} = 1,2$, we find that the two elements in the second row contain $\Psi(\underline{X}_i)$ and $\Psi'(\underline{X}_i)$. We proceed to simplify the expressions for these matrix elements. We use the fact that the functions $\underline{X}_i$ satisfy (5.1.19). It follows that

$$\Psi(X_i) = \frac{2X_i}{T_2(T_1-X_i)}, \quad i = 1,2. \tag{5.1.85}$$

Furthermore, it follows from (5.1.58) that $\Psi'(\underline{x})$ can be expressed in terms of $\Psi(\underline{x})$ and $\underline{h}(\underline{x})$ as follows:

$$\Psi'(x) = \frac{h(x)\Psi(x)}{x(h(x)-1)}, \quad x > 0. \tag{5.1.86}$$

By putting $\underline{x} = \underline{X}_i$ and using (5.1.85) we get

$$\Psi'(X_i) = \frac{2h(X_i)}{T_2(T_1-X_i)(h(X_i)-1)}. \tag{5.1.87}$$

By now applying (5.1.85) and (5.1.87) we find that the Jacobian matrix evaluated at $\underline{z} = \bar{z}^i$, $\underline{i} = 1,2$, can be written in the following form:

$$\frac{\partial f}{\partial z}(\bar{z}^i) = \begin{pmatrix} -\mu_1 & \mu_1 \\ \dfrac{\mu_2 h(X_i)}{h(X_i)-1} & -\dfrac{\mu_2 T_1}{T_1-X_i} \end{pmatrix}, \quad i = 1,2. \tag{5.1.88}$$

The eigenvalues of this matrix satisfy the equation

$$\lambda^2 + \left(\mu_1 + \frac{\mu_2 T_1}{T_1-X_i}\right)\lambda + \frac{\mu_1\mu_2 B}{(T_1-X_i)(h(X_i)-1)} = 0, \tag{5.1.89}$$

where the definition (5.1.25) of the function $\bar{\underline{T}}_1$ can be used to write $\underline{B}$ in the form

$$B = X_i h(X_i) - T_1 = \bar{T}_1(X_i) - T_1. \tag{5.1.90}$$

The inequalities $0 < \underline{X}_1 < \underline{X}_2 < \underline{T}_1$ show that the coefficient of λ in (5.1.89) is positive. Furthermore, the inequalities for Ψ' and $\underline{h}$ in (2.4.25) and (5.1.31) show that the constant term in (5.1.89) has the same sign as $\underline{B}$. We note from (5.1.38) that $\underline{\bar{T}}_1$ is an increasing function. We note furthermore from (5.1.79), (5.1.80) that

$$X_1(T_1,T_2) \leq \bar{X}_1(T_1) \leq X_2(T_1,T_2), \tag{5.1.91}$$

where equality signs hold on the threshold, while the inequalities are strict everywhere above the threshold. Above the threshold, therefore, we conclude that

$$\bar{T}_1(X_1) < \bar{T}_1(\bar{X}_1(T_1)) = T_1 \tag{5.1.92}$$

and

$$\bar{T}_1(X_2) > \bar{T}_1(\bar{X}_1(T_1)) = T_1. \tag{5.1.93}$$

We conclude that $\underline{B}$ is negative at the critical point $\underline{\bar{z}}^1$ everywhere above the threshold. It follows from criterion (AIV.17) that $\underline{\bar{z}}^1$ is unstable above the threshold. Furthermore, $\underline{B}$ is positive at the critical point $\underline{\bar{z}}^2$ above the threshold. Criterion (AIV.17) allows us to conclude that the critical point $\underline{\bar{z}}^2$ is asymptotically stable above the threshold.

The function $\underline{X}_1(\underline{T}_1,\underline{T}_2)$ is shown dashed in Figures 5.1C and 5.1D to indicate that it corresponds to an unstable critical point. The critical points corresponding to the worm loads 0 and $\underline{X}_2(\underline{T}_1,\underline{T}_2)$ are asymptotically stable. The two functions $\underline{X}_1$ and $\underline{X}_2$ take the same values on the threshold where $\underline{T}_1 = \underline{\bar{\bar{T}}}_1(\underline{T}_2)$; the worm load is then equal to $\underline{\bar{X}}_1(\underline{T}_1) = \underline{\bar{X}}_2(\underline{T}_2)$. Our analysis allows no conclusion about the stability of the corresponding critical point.

5.1.4 Global Analysis

Phase portraits for the system of differential equations (5.1.11), (5.1.12) are shown below and above threshold in Figures 5.1E and 5.1F. The phase portraits can be used to show that the rectangle $\underline{R}$ is positively invariant. They can also be used to show that the solution approaches a critical point for any initial point in the rectangle $\underline{R}$. Thus every solution approaches the origin if in parameter space the point $(\underline{T}_1, \underline{T}_2)$ lies below the threshold. This corresponds epidemiologically to the disappearance of infection from the community. Above the threshold, three possibilities exist, corresponding to the three critical points. The critical point with worm load $\underline{X}_1(\underline{T}_1, \underline{T}_2)$ is unstable. Its occurrence in an epidemiological situation can therefore be ruled out; small disturbances of this equilibrium solution will cause the solution to approach one of the stable critical points. This unstable critical point has been termed "breakpoint" by Macdonald (1965).

The origin and the critical point with worm load $\underline{X}_2(\underline{T}_1, \underline{T}_2)$ are both asymptotically stable. The domain of attraction of the origin is small if the worm load $\underline{X}_1(\underline{T}_1, \underline{T}_2)$ at the breakpoint is small. We have shown that $\underline{X}_1$ is a decreasing function of both $\underline{T}_1$ and $\underline{T}_2$. Thus we expect the domain of attraction of the origin to be small if $\underline{T}_1$ and/or $\underline{T}_2$ is large. If a community has an endemic infection level with a positive worm load, then we can conclude that the community lies above the threshold and that the infection level is determined by the critical point with worm load $\underline{X}_2(\underline{T}_1, \underline{T}_2)$.

5.1.5 Modes of Eradication

The schistosomiasis model indicates that there are two conceptually different modes of eradication at our disposal. Eradication can be achieved either by a permanent reduction of the values of the transmission factors $\underline{T}_1$ and/or $\underline{T}_2$ to below the threshold, or by a momentary reduction of the state variables $\underline{z}_1$ and/or $\underline{z}_2$ to the domain of attraction of the origin. The amount of reduction required for eradication in the latter case is to a large extent determined by the breakpoint.

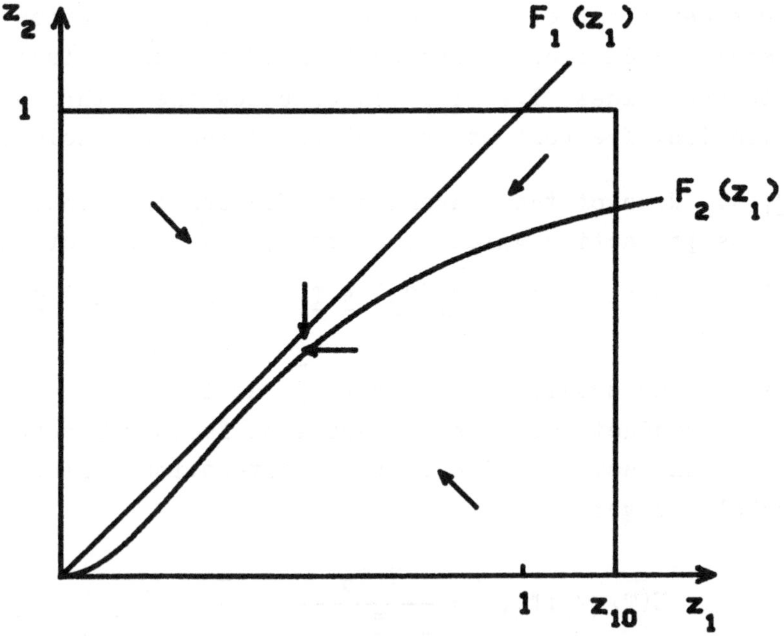

<u>Figure 5.1E.</u> Phase portrait for the schistosomiasis model below the threshold.

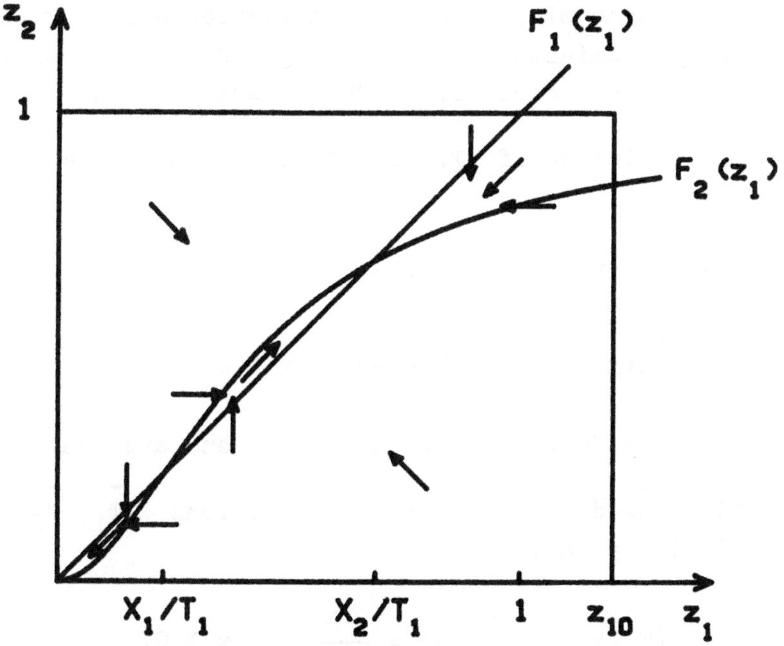

<u>Figure 5.1F.</u> Phase portrait for the schistosomiasis model above the threshold.

We consider first eradication of an endemic infection based on the threshold. Reduction of one or both of the transmission factors $\underline{T}_1$, $\underline{T}_2$ to new values $\underline{T}_{11}$, $\underline{T}_{21}$ on or below the threshold leads to eradication. The post-action values on the threshold satisfy $\underline{T}_{21} = \bar{\bar{\underline{T}}}_2(\underline{T}_{11})$. None of the transmission factors is allowed to increase from its pre-action value. Therefore, the post-action values are found in the intervals $\bar{\bar{\underline{T}}}_1(\underline{T}_2) \leq \underline{T}_{11} \leq \underline{T}_1$, $\bar{\bar{\underline{T}}}_2(\underline{T}_1) \leq \underline{T}_{21} \leq \underline{T}_2$.

We define the eradication effort $\bar{\underline{E}}$ as the ratio $\underline{T}_1\underline{T}_2/(\underline{T}_{11}\underline{T}_{21})$ of of the product of the transmission factors before and after the eradication action. Since the post-action values lie on the threshold we get

$$\bar{E}(T_1,T_2;T_{11}) = \frac{T_1T_2}{T_{11}\bar{\bar{T}}_2(T_{11})}, \quad \bar{\bar{T}}_1(T_2) \leq T_{11} \leq T_1. \qquad (5.1.94)$$

For given pre-action values $\underline{T}_1,\underline{T}_2$ we shall find that post-action value $\underline{T}_{11}$ that minimizes the eradication effort $\bar{E}$. We shall prove that the denominator of the right-hand side of (5.1.94) is a decreasing function of $\underline{T}_{11}$. This implies that the eradication effort is minimized if $\underline{T}_1$ is reduced to the value $\bar{\bar{\underline{T}}}_1(\underline{T}_2)$ and $\underline{T}_2$ is held fixed.

We note first that the relation

$$\bar{T}_2(x) = -2\,\frac{d}{dx}\left(\frac{x}{\Psi(x)}\right), \qquad x > 0, \qquad (5.1.95)$$

follows from differentiation of $\underline{x}/\Psi(\underline{x})$ and application of (5.1.58), (5.1.26). The definitions of the functions $\bar{\underline{T}}_1$, $\bar{\underline{T}}_2$ in (5.1.25), (5.1.26) imply that

$$(\bar{T}_1(x) - x)\bar{T}_2(x) = \frac{2x}{\Psi(x)}, \qquad x > 0. \qquad (5.1.96)$$

Differentiation of this expression and application of (5.1.95) shows that the ratio of the derivatives of $\bar{T}_1$ and $\bar{T}_2$ can be expressed as follows:

$$\frac{\bar{T}_2'(x)}{\bar{T}_1'(x)} = -\frac{\bar{T}_2(x)}{\bar{T}_1(x)-x}, \quad x > 0. \tag{5.1.97}$$

Since $\bar{\underline{X}}_1$ is the inverse of $\bar{\underline{T}}_1$, we get

$$\bar{X}_1'(T_1) = \frac{1}{\bar{T}_1'(\bar{X}_1(T_1))}, \quad T_1 > 0. \tag{5.1.98}$$

Differentiation of the function $\bar{\bar{\underline{T}}}_2$ defined in (5.1.28) and application of (5.1.97), (5.1.98) leads to the following expression for the derivative of the threshold function $\bar{\bar{\underline{T}}}_2$:

$$\bar{\bar{T}}_2'(T_1) = -\frac{\bar{\bar{T}}_2(T_1)}{T_1-\bar{X}_1(T_1)}, \quad T_1 > 0. \tag{5.1.99}$$

It can now be proved that

$$\frac{d}{dT_1}(T_1\bar{\bar{T}}_2(T_1)) = \bar{\bar{T}}_2'(T_1)\bar{X}_1(T_1) < 0, \quad T_1 > 0. \tag{5.1.100}$$

The proof of this relation follows after differentiation and a two-fold application of (5.1.99).

We have thus proved that the eradication effort is minimized if $\underline{T}_1$ is reduced to its threshold value $\bar{\bar{\underline{T}}}_1(\underline{T}_2)$ and $\underline{T}_2$ is held fixed. The minimum eradication effort $\underline{E}$ is therefore defined on and above the threshold by the relation

$$E(T_1,T_2) = \frac{T_1}{\bar{\bar{T}}_1(T_2)}. \tag{5.1.101}$$

Our results lead to the recommendation that eradication actions using the threshold should be directed at sufficiently large reductions of the transmission factor $\underline{T}_1$. It is however recognized that the recommendation may be reversed if cost data show that reductions of $\underline{T}_2$ are sufficiently less expensive than reductions of $\underline{T}_1$. The minimum eradication effort $\underline{E}$ is for each community a useful measure of the amount of parameter modification required for eradication. It quantifies the necessary perturbation in parameter space.

The second mode of eradication is based on perturbation in the state space. A simultaneous momentary reduction of both the state variables $\underline{z}_1$ and $\underline{z}_2$ to a point in the domain of attraction of the origin is predicted by the model to lead to eradication. By referring to Figure 5.1F and recalling that $\underline{z}_1 = \underline{x}_1/\underline{T}_1$ and $\underline{z}_2 = \underline{p}_2$, we find that a reduction of $\underline{x}_1$ to below the breakpoint worm load $\underline{X}_1(\underline{T}_1,\underline{T}_2)$ and a simultaneous reduction of $\underline{p}_2$ to below $\underline{X}_1(\underline{T}_1,\underline{T}_2)/\underline{T}_1$ will suffice for eradication.

Reductions of the worm load can in practice be achieved by mass administration of antischistosomal drugs. If such a chemotherapy campaign alone is to lead to eradication, then its efficiency, as measured by the proportionate reduction of the worm load in the community, must clearly exceed the ratio

$$\bar{R}(T_1,T_2) = 1 - \frac{X_1(T_1,T_2)}{X_2(T_1,T_2)} \quad . \qquad (5.1.102)$$

The efficiency of a chemotherapy campaign is always less than 100%, and even 90% may be hard to reach. Limitations are caused by the difficulty of identifying and getting cooperation from all individuals in the community, and also by the inefficiency of available drugs.

Formula (5.1.102) can be used to compute the lower bound $\bar{\underline{R}}$ of the required chemotherapy campaign efficiency for any community for which the values of the transmission factors $\underline{T}_1$ and $\underline{T}_2$ are known. Now individual values of the transmission factors are less informative than the minimum eradication effort $\underline{E}$. We proceed therefore to study $\bar{\underline{R}}$ as a function of $\underline{T}_1$ and $\underline{E}$. From (5.1.101) we find that $\underline{T}_2$ is determined as a function of $\underline{T}_1$ and $\underline{E}$ by the relation

$$T_2 = \bar{\bar{T}}_2(T_1/E). \qquad (5.1.103)$$

The lower bound of the required chemotherapy campaign efficiency can therefore be written

$$R(T_1,E) = \bar{R}(T_1,\bar{\bar{T}}_2(T_1/E)) = 1 - \frac{X_1(T_1,\bar{\bar{T}}_2(T_1/E))}{X_2(T_1,\bar{\bar{T}}_2(T_1/E))} . \tag{5.1.104}$$

We conjecture that $\underline{R}$ is a monotonically increasing function of $\underline{T}_1$ for each $\underline{E} > 1$. The analytic proof of this conjecture is left as an open problem. The conjecture is well supported by numerical evaluations. The conjecture implies that $\underline{R}(\underline{T}_1,\underline{E})$ is larger than its limiting value as $\underline{T}_1$ approaches zero. In order to find this limiting value, we evaluate first the asymptotic expression for the ratio $\underline{X}_1/\underline{X}_2$ in the right-hand side of (5.1.104). Because of the inequalities $0 < \underline{X}_1 < \underline{X}_2 < \underline{T}_1$ we conclude that both $\underline{X}_1$ and $\underline{X}_2$ approach zero. By now using the defining relation (5.1.22), which is satisfied by both $\underline{X}_1$ and $\underline{X}_2$, the definition (5.1.20) of the function $\hat{T}_1$, and the asymptotic expressions (2.4.29) and (5.1.55) for Ψ and $\bar{\bar{T}}_2$, we find that

$$T_1 \sim x + \frac{T_1^2}{4E^2x} , \quad T_1 \to 0. \tag{5.1.105}$$

By solving for $\underline{x}$, we get

$$x \sim \frac{T_1}{2} (1 \pm (E^2-1)^{1/2}/E), \quad T_1 \to 0, \quad E \geq 1. \tag{5.1.106}$$

In both of these relations, $\underline{x}$ stands for $\underline{X}_i(\underline{T}_1,\bar{\bar{\underline{T}}}_2(\underline{T}_1/\underline{E}))$, $\underline{i}=1,2$. The + sign in (5.1.106) gives the asymptotic expression for $\underline{X}_2$ and the - sign gives the expression for $\underline{X}_1$. By inserting these asymptotic expressions into (5.1.104), we find that

$$R(T_1,E) > R_0(E) = \frac{2(E^2-1)^{1/2}}{E+(E^2-1)^{1/2}}, \; T_1 > 0, \; E \geq 1. \tag{5.1.107}$$

For each value $\underline{E} \geq 1$ of the minimum eradication effort, $\underline{R}_0(\underline{E})$ gives a lower bound for the required chemotherapy campaign efficiency. This lower bound is valid for all communities described by the fixed value of $\underline{E}$ since the inequality in (5.1.107) holds for all $\underline{T}_1 > 0$.

Conclusions regarding the feasibility of eradicating schistosomiasis through a chemotherapy campaign can be drawn on the basis of the entries in Table 5.1. If the minimum eradication effort exceeds ten (a moderately large value), then eradication by use of the breakpoint requires that the efficiency of the chemotherapy campaign be larger than 99.75%. This value is impossible to achieve in paractice and therefore we conclude that eradication using the breakpoint is impossible in a community with $\underline{E}$ larger than ten. Indeed, if an efficiency of 90% can be reached, than Table 5.1 shows that eradication by use of the breakpoint is not possible unless the minimum eradication effort $\underline{E}$ is less than 1.74, which means that the community is already very close to the threshold. These questions are discussed in more detail by Nåsell (1984).

Table 5.1 The lower bound $\underline{R}_0(\underline{E})$ of the required chemotherapy campaign efficiency is listed as function of the minimum eradication effort $\underline{E}$.

E	$R_0(E)$
1.74	0.9
2	0.928
5.05	0.99
10	0.9975
100	0.999975

5.1.6 Epidemiological Measures in the Endemic Case

As for previous models there are a number of epidemiological measures that can be used as indicators of the infection level in a community above threshold, i.e. with $\underline{T}_1 > \underline{\underline{T}}_1(\underline{T}_2)$. As before we confine ourselves to a consideration of such epidemiological measures in the steady state.

The steady-state expected number of parasites per human host (the worm load) is given by $\underline{X}_2(\underline{T}_1,\underline{T}_2)$, and the steady-state expected number of mated female parasites per human host is given by

$$\frac{1}{2}\Psi(X_2(T_1,T_2)) = \frac{X_2(T_1,T_2)}{(T_1-X_2(T_1,T_2))T_2} . \tag{5.1.108}$$

The expression on the right-hand side of (5.1.108) is found from the fact that $\underline{X}_2$ satisfies (5.1.23). The steady-state infection probability $\underline{P}_2$ among the snails is equal to

$$P_2(T_1,T_2) = \frac{X_2(T_1,T_2)}{T_1} . \tag{5.1.109}$$

The steady-state human infection rate $\underline{H}_1$ (the public health factor) depends on the three parameters $\underline{u}_1$, μ_1, $\underline{T}_2$, where $\underline{u}_1$ is defined by (4.1.34). It is found from (5.1.3) and (5.1.7) that $\underline{H}_1$ is equal to

$$H_1(u_1,\mu_1,T_2) = u_1 P_2(u_1/\mu_1,\ T_2) = \mu_1\ X_2(u_1/\mu_1,T_2). \tag{5.1.110}$$

By inserting this expression into (2.4.40) we find the following expression for the steady-state probability $\underline{p}_{FO}(\underline{s})$ of any host of age $\underline{s}$ to be free of female parasites:

$$p_{FO}(s) = \exp\left(-\frac{1}{2}X_2(u_1/\mu_1,T_2)(1-e^{-\mu_1 s})\right), \quad s \geq 0. \tag{5.1.111}$$

Explicit expressions for prevalence, incidence and recovery probability in steady state are found by inserting this relation into (2.4.32), (2.4.39), and (2.4.36), respectively.

5.1.7 Control Efficiencies

All epidemiological quantities treated in the previous subsection can be used to judge the effect of a control action. We confine ourselves to a study of only two control efficiency functions, namely those that measure the efficiencies of control of the steady-state worm load $\underline{X}_2(\underline{T}_1,\underline{T}_2)$ through reductions of $\underline{T}_1$ and

$\underline{T}_2$, respectively. The two control efficiency functions are denoted $\underline{C}_1$ and $\underline{C}_2$. They are defined as follows:

$$C_1(T_1,T_2) = \frac{T_1}{X_2}\frac{\partial X_2}{\partial T_1} , \qquad (5.1.112)$$

$$C_2(T_1,T_2) = \frac{T_2}{X_2}\frac{\partial X_2}{\partial T_2} . \qquad (5.1.113)$$

The arguments of the function $\underline{X}_2$ are deleted in these expressions. Both control efficiency functions are defined above the threshold. The partial derivatives of the function $\underline{X}_2$ are readily evaluated since $\underline{X}_2(\underline{T}_1,\underline{T}_2)$ satisfies (5.1.22) and (5.1.23). Thus we find

$$\frac{\partial X_2}{\partial T_1} = \frac{1}{\frac{\partial \hat{T}_1}{\partial x}(X_2;T_2)} \qquad (5.1.114)$$

and

$$\frac{\partial X_2}{\partial T_2} = \frac{1}{\frac{\partial \hat{T}_2}{\partial x}(X_2;T_1)} . \qquad (5.1.115)$$

From the expression for the partial derivative of $\hat{\underline{T}}_1$ with respect to $\underline{x}$ in (5.1.60) we find that

$$\frac{\partial X_2}{\partial T_1} = \frac{T_2}{T_2-\bar{T}_2(X_2)} . \qquad (5.1.116)$$

The expression for the partial derivative of $\hat{\underline{T}}_2$ with respect to $\underline{x}$ in (5.1.61) gives

$$\frac{\partial \hat{T}_2}{\partial x}(X_2;T_1) = \frac{\bar{T}_2(X_2)(\bar{T}_1(X_2) - T_1)}{(T_1 - X_2)^2} . \qquad (5.1.117)$$

This expression can be simplified. We make use of the fact that the argument of both $\bar{\underline{T}}_1$ and $\bar{\underline{T}}_2$ is $\underline{X}_2$.

From the definition of the function $\bar{\underline{T}}_1$ in (5.1.25) we find that

$$\bar{T}_1(X_2) = X_2 h(X_2). \tag{5.1.118}$$

By using (5.1.108) and the definition of the function $\bar{\underline{T}}_2$ in (5.1.26) we get

$$\bar{T}_2(X_2) = \frac{(T_1 - X_2)T_2}{(h(X_2) - 1)X_2} . \tag{5.1.119}$$

By eliminating $\underline{h}(\underline{X}_2)$ from (5.1.118) and (5.1.119) we find that $\bar{\underline{T}}_1(\underline{X}_2)$ and $\bar{\underline{T}}_2(\underline{X}_2)$ are related as follows:

$$\bar{T}_1(X_2) = X_2 + \frac{(T_1 - X_2)T_2}{\bar{T}_2(X_2)} . \tag{5.1.120}$$

By inserting this expression into (5.1.117) and using (5.1.115) we find that the partial derivative of $\underline{X}_2$ with respect to $\underline{T}_2$ can be written

$$\frac{\partial X_2}{\partial T_2} = \frac{T_1 - X_2}{T_2 - \bar{T}_2(X_2)} . \tag{5.1.121}$$

The two control efficiency functions defined by (5.1.112),(5.1.113) can now be written in the following form:

$$C_1 = \frac{T_1 T_2}{(T_2 - \bar{T}_2(X_2))X_2} , \tag{5.1.122}$$

$$C_2 = \frac{(T_1 - X_2)T_2}{(T_2 - \bar{T}_2(X_2))X_2} . \tag{5.1.123}$$

The inequalities $\underline{X}_2 < \underline{T}_1$ and $0 < \bar{\underline{T}}_2(\underline{X}_2) < \underline{T}_2$ allow us to conclude that $\underline{C}_1 > 1$. Thus, the efficiency of control of the steady-state worm load $\underline{X}_2$ through reductions of $\underline{T}_1$ is always larger than one. The ratio of the two control efficiency functions is equal to

$$\frac{C_1}{C_2} = \frac{T_1}{T_1 - X_2} = \frac{1}{1 - P_2} > 1. \qquad (5.1.124)$$

Thus it is always more efficient to control the steady-state worm load through a reduction of the transmission factor $\underline{T}_1$ than through a reduction of the transmission factor $\underline{T}_2$. The advantage is substantial if the steady-state snail prevalence $\underline{P}_2$ is close to one, but it is small if $\underline{P}_2$ is small.

Studies of efficiencies of control of other epidemiological measures can be carried out, and a number of qualitative results similar to (5.1.124) can be established. We omit details.

5.2 The Macdonald Schistosomiasis Model

The path-breaking schistosomiasis transmission model presented by Macdonald (1965) accounts for superinfection of human beings and for monogamous mating between male and female parasites, and also for latency of the infection in the snail population. We treat it here by adding a hypothesis of snail latency to the basic schistosomiasis model dealt with in the preceeding section. We use a model formulation that is stochastic for the human phase of the life cycle and deterministic for the snail phase.

The model takes the explicit form of a collection of $2\underline{N}_1$ Markov chains and three deterministic functions as follows:

$$\{F^{(1)}(t), M^{(1)}(t), \ldots, F^{(N_1)}(t), M^{(N_1)}(t), S_2(t), L_2(t), I_2(t)\} , \quad t \geq 0.$$

To each human host $\underline{k}$ there correspond two Markov chains $\underline{F}^{(k)}$ and $\underline{M}^{(k)}$ that count the number of female and male parasites in the host. The snail phase of the life cycle is represented in the model by the three functions $\underline{S}_2$, $\underline{L}_2$, and $\underline{I}_2$, whose values at time $\underline{t}$ are interpreted as the number of susceptible, latent, and infected snails, respectively. It is not clear from Macdonald's paper if he assumes that there is a constant rate at which latent snails become infected or if he assumes that the latent period is constant. We base our model on the former assumption and introduce $\underline{k}_2$ to denote the rate at which each latent snail becomes infected. All snails are assumed to have the same death rate μ_2.

The arguments of the preceding section show that the expected number of parasites per human host at time $\underline{t}$, $\underline{x}_1(\underline{t})$, satisfies the differential equation

$$x_1' = h_1(t) - \mu_1 x_1. \tag{5.2.1}$$

For the snail phase we find from equations (2.7.1)-(2.7.3) that

$$S_2' = \mu_2 N_2 - h_2(t) S_2 - \mu_2 S_2, \tag{5.2.2}$$

$$L_2' = h_2(t) S_2 - k_2 L_2 - \mu_2 L_2, \tag{5.2.3}$$

$$I_2' = k_2 L_2 - \mu_2 I_2. \tag{5.2.4}$$

Under the assumption that the total initial number of snails is equal to $\underline{N}_2$, we conclude from these equations that the total number of snails is $\underline{N}_2$ at each $\underline{t} \geq 0$. Thus, the relation

$$S_2(t) = N_2 - L_2(t) - I_2(t), \quad t \geq 0, \tag{5.2.5}$$

can be used to reduce the number of equations by one.

The infection rates $\underline{h}_1(\underline{t})$ and $\underline{h}_2(\underline{t})$ are found by arguments very similar to those in Section 5.1. The results are

$$h_1(t) = \pi_1 \lambda_2 I_2(t) \tag{5.2.6}$$

and

$$h_2(t) = \tfrac{1}{2} \pi_2 \lambda_1 N_1 \Psi(x_1(t)). \tag{5.2.7}$$

By inserting relations (5.2.5)-(5.2.7) into equations (5.2.1), (5.2.3), (5.2.4) we get the following system of differential equations for the three state variables $\underline{x}_1$, $\underline{L}_2$, $\underline{I}_2$:

$$x_1' = \pi_1\lambda_2 I_2 - \mu_1 x_1, \tag{5.2.8}$$

$$L_2' = \frac{1}{2}\pi_2\lambda_1 N_1 \Psi(x_1)(N_2-L_2-I_2) - (k_2+\mu_2)L_2, \tag{5.2.9}$$

$$I_2' = k_2 L_2 - \mu_2 I_2. \tag{5.2.10}$$

The first step in the treatment of these equations is to search for suitable transmission factors. These are found in a simple way by investigating the critical points of the system of equations (5.2.8)-(5.2.10) and then using the results of the preceding section. We denote any critical point by $(\bar{x}_1, \bar{L}_2, \bar{I}_2)$. By putting the right-hand sides of the three equations equal to zero, we find that $\bar{x}_1$ satifies the relation

$$\frac{\pi_2\lambda_1 N_1}{\mu_2}\Psi(\bar{x}_1)\left(\frac{k_2\pi_1\lambda_2 N_2}{(k_2+\mu_2)\mu_1} - \bar{x}_1\right) = 2\bar{x}_1. \tag{5.2.11}$$

This relation has exactly the same form as (5.1.19) if we define two transmission factors T_1 and T_2 by

$$T_1 = \frac{k_2\pi_1\lambda_2 N_2}{(k_2+\mu_2)\mu_1}, \tag{5.2.12}$$

$$T_2 = \frac{\pi_2\lambda_1 N_1}{\mu_2}. \tag{5.2.13}$$

A comparison with the transmission factors defined in the previous section by (5.1.7) and (5.1.8) shows that latency in the snail population can be accounted for by identifying the number of snails that are "effective" in transmitting schistosome infections. This number is equal to the total number of snails N_2 multiplied by the the factor $k_2/(k_2+\mu_2)$, which is the proportion of snails that survive the latent period in steady state.

The quasi-dimension of the new parameter k_2 is T^{-1}. It follows that the quasi-dimension of T_1 defined by (5.2.12) is $P\,H^{-1}$, i.e. the same as the quasi-dimension of x_1. Thus we can introduce a new state variable z_1 free of quasi-dimension by setting

$$z_1 = x_1/T_1. \tag{5.2.14}$$

The differential equation for $\underline{z}_1$ takes the simple form

$$z_1' = \mu_1(z_3 - z_1) \tag{5.2.15}$$

if we agree to define

$$z_3 = \frac{(k_2 + \mu_2)I_2}{k_2 N_2} . \tag{5.2.16}$$

Thus, $\underline{z}_3$ is the number of infected snails divided by the number of snails that are "effective" in transmitting infections. As a third state variable free of quasi-dimension we introduce

$$z_2 = (L_2 + I_2)/N_2. \tag{5.2.17}$$

This choice of $\underline{z}_2$ leads to slightly simpler differential equations than if we define $\underline{z}_2$ to be the proportion of snails in the latent state. It follows that the state variables $\underline{z}_2$ and $\underline{z}_3$ satisfy the equations

$$z_2' = \mu_2(\tfrac{1}{2} T_2\Psi(T_1 z_1)(1 - z_2) - z_2), \tag{5.2.18}$$

$$z_3' = (k_2 + \mu_2)(z_2 - z_3). \tag{5.2.19}$$

Any critical point of the system of equations (5.2.15), (5.2.18), (5.2.19) is denoted by $\bar{z} = (\bar{z}_1, \bar{z}_2, \bar{z}_3)$. The components of the critical point are found to satisfy the following relations:

$$T_2\Psi(T_1\bar{z}_1)(1 - \bar{z}_1) = 2\,\bar{z}_1 \tag{5.2.20}$$

$$\bar{z}_2 = \bar{z}_3 = \bar{z}_1. \tag{5.2.21}$$

Since equation (5.2.20) is identical in form to equation (5.1.17), we can use the results in Section 5.1 to draw the following conclusions about the critical points of the system of equations (5.2.15), (5.2.18), (5.2.19): There is a threshold function in $\underline{T}_1$-$\underline{T}_2$-space, given by $\underline{T}_1 = \overset{=}{\underline{T}}_1(\underline{T}_2)$. Below the threshold there is only one critical point, namely the origin. Above the threshold there are two additional critical points, $\underline{\bar{z}}^1$ and $\underline{\bar{z}}^2$. Each component of $\bar{z}^i$, $i = 1,2$, is equal to $X_i(T_1,T_2)/T_1$. The two critical points $\bar{z}^1$ and $\bar{z}^2$ coincide on the threshold.

We proceed to analyze the local stability of the critical points. The system of equations (5.2.15), (5.2.18), (5.2.19) is written $\underline{z}' = \underline{f}(\underline{z})$ in vector form. The Jacobian matrix is found to be equal to

$$\frac{\partial f}{\partial z}(z) = \begin{pmatrix} -\mu_1 & 0 & \mu_1 \\ \frac{1}{2}\mu_2 T_1 T_2 \Psi'(T_1 z_1)(1-z_2) & -\mu_2(1+\frac{1}{2}T_2\Psi(T_1 z_1)) & 0 \\ 0 & k_2+\mu_2 & -k_2-\mu_2 \end{pmatrix}. \tag{5.2.22}$$

Evaluating the Jacobian matrix at the origin, the critical point that corresponds to absence of infection from the community, we find that

$$\frac{\partial f}{\partial z}(0) = \begin{pmatrix} -\mu_1 & 0 & \mu_1 \\ 0 & -\mu_2 & 0 \\ 0 & k_2+\mu_2 & -k_2-\mu_2 \end{pmatrix}. \tag{5.2.23}$$

This matrix has the three eigenvalues $-\mu_1$, $-\mu_2$, and $-k_2-\mu_2$. They are all negative in the case we are considering, namely with all basic parameters positive. We conclude (see Appendix IV) that the origin is always an asymptotically stable critical point.

By using (5.1.85) and (5.1.87) we find that the Jacobian matrix evaluated at $\bar{z}^i$, $i = 1,2$, can be written in the following form:

$$\frac{\partial f}{\partial z}(\bar{z}^i) = \begin{pmatrix} -\mu_1 & 0 & \mu_1 \\ \dfrac{\mu_2 h(X_i)}{h(X_i)-1} & -\dfrac{\mu_2 T_1}{T_1-X_i} & 0 \\ 0 & k_2+\mu_2 & -k_2-\mu_2 \end{pmatrix}. \tag{5.2.24}$$

The eigenvalues of this matrix satisfy the equation

$$\lambda^3 + A\lambda^2 + B\lambda + C = 0, \tag{5.2.25}$$

where

$$A = \mu_1 + \mu_2 + k_2 + \frac{\mu_2 T_1}{T_1 - X_i}, \tag{5.2.26}$$

$$B = \mu_1(\mu_2 + k_2) + \frac{(\mu_1 + \mu_2 + k_2)\mu_2 T_2}{T_1 - X_i}, \tag{5.2.27}$$

and

$$C = \frac{\mu_1 \mu_2(\mu_2 + k_2)}{(T_1 - X_i)(h(X_i) - 1)} (X_i h(X_i) - T_1). \tag{5.2.28}$$

It is readily verified that $\underline{A} > 0$ and that $\underline{B} > 0$. Furthermore, by deleting three positive terms we find that

$$AB - C > \frac{\mu_1 \mu_2(\mu_2 + k_2)}{T_1 - X_i} \left[T_1 - \frac{X_i h(X_i) - T_1}{h(X_i) - 1} \right]$$

$$= \frac{\mu_1 \mu_2(\mu_2 + k_2) h(X_i)}{h(X_i) - 1} > 0. \tag{5.2.29}$$

Finally, by invoking the definition (5.1.25) of the function $\bar{\underline{T}}_1$ and the inequalities (5.1.92),(5.1.93), we conclude from (5.2.28) that $\underline{C}$ is negative for $\underline{i} = 1$ and positive for $\underline{i} = 2$. By using the criterion (AIV.19) we conclude that all eigenvalues of the matrix $\frac{\partial f}{\partial z}(\bar{\underline{z}}^2)$ have negative real parts and that at least one eigenvalue of the matrix $\frac{\partial f}{\partial z}(\bar{\underline{z}}^1)$ has positive real part. It follows that the critical point $\bar{\underline{z}}^2$ is asymptotically stable above the threshold and that the critical point $\bar{\underline{z}}^1$ is unstable above the threshold.

These local stability results all agree with the corresponding results established for the schistosomiasis model without latency in Section 5.1. However, it still remains to establish global results that describe the asymptotic behaviour of the solution of the system of equations (5.2.15), (5.2.18), (5.2.19) for arbitrary (but biologically meaningful) initial values, and not only for initial values sufficiently close to the critical points. For this purpose we shall apply some powerful general results recently established by Hirsch (1984).

The results of Hirsch are formulated in the framework of dynamical systems. We consider the system of differential equations $\underline{z}' = \underline{f}(\underline{z})$ with initial value $\underline{z}(0) = \underline{v}$, where $\underline{z}$ and $\underline{v}$ are $\underline{n}$-dimensional vectors. We assume that the Jacobian matrix $\frac{\partial f}{\partial z}$ is continuous on some open subset $\underline{X}$ of $\underline{R}^n$. It follows then that the initial value problem has a unique solution $\underline{z}(\underline{t},\underline{v})$ on some $\underline{t}$-interval for each $\underline{v}$ in $\underline{X}$. It is common to think of this solution as a mapping from the $\underline{t}$-interval on which it exists into the set $\underline{X}$ for each fixed initial value $\underline{v}$. However, the approach in dynamical systems is the opposite. We can clearly use the solution to define a mapping, for each fixed value of $\underline{t}$, of the set of initial values into the solution values in $\underline{X}$. The mapping is denoted by ϕ_t. It is defined on the subset of initial values in $\underline{X}$ for which solutions exist at time $\underline{t}$. For any $\underline{v}$ in this subset we write

$$\phi_t(v) = z(t,v). \tag{5.2.30}$$

The collection $\phi = \{\phi_t\}$ of such maps for $\underline{t} \geq 0$ is called a <u>flow</u>.

In order to be able to formulate Hirsch's result, we need an additional concept. We consider two vectors $\underline{x}$ and $\underline{y}$ in $\underline{R}^n$ with components $\underline{x}_i$ and $\underline{y}_i$, respectively ($\underline{i} = 1,2,\ldots,\underline{n}$). If each component of $\underline{x}$ is smaller than or equal to the corresponding component of $\underline{y}$, but with strict inequality for at least one $\underline{i}$, then we write $\underline{x} < \underline{y}$. If each component of $\underline{x}$ is strictly smaller then the corresponding component of $\underline{y}$, then we write $\underline{x} << \underline{y}$. Thus,

$$\begin{aligned} &x < y \text{ if } x_i \leq y_i,\ i=1,\ldots,n \text{ and } x \neq y \\ &x << y \text{ if } x_i < y_i,\ i=1,\ldots,n. \end{aligned} \tag{5.2.31}$$

A map $\underline{F}$ from the subset $\underline{X}$ into $\underline{R}^n$ is called strongly monotone if $\underline{x} < \underline{y}$ implies $\underline{F}(\underline{x}) << \underline{F}(\underline{y})$. The flow ϕ is called <u>strongly monotone</u> if the map ϕ_t is strongly monotone for each $\underline{t} > 0$.

In the sequel we use $\underline{V}$ to denote a compact positively invariant subset of $\underline{X}$ with the property that all biologically meaningful initial values $\underline{v}$ belong to $\underline{V}$. The solution $\underline{z}(\underline{t},\underline{v})$, with $\underline{v}$ in the set $\underline{V}$, will then exist for all $\underline{t} \geq 0$, and the orbit through each $\underline{v}$ will be a subset of $\underline{V}$. A point $\underline{v}$ is called convergent if the solution $\underline{z}(\underline{t},\underline{v})$ with initial value $\underline{v}$ approaches a stationary point as time aproaches infinity. The first result of Hirsch that we shall apply can now be formulated as follows:

Theorem 1. Let the set $\underline{V}$ be compact and positively invariant and let the number of stationary points be finite. Assume that the flow ϕ is strongly monotone. Then the set of nonconvergent points in $\underline{V}$ has measure zero.

If the hypotheses of the theorem are satisfied, then we can conclude that the only biologially meaningful solutions of the system of differential equations are those that approach a stationary point as time approaches infinity.

In order to be able to apply this theorem, we need a criterion for determining when the flow ϕ has the important property of being strongly monotone. Sufficient conditions for this to occur are also given by Hirsch (1984). We define the vector field $\underline{f}$ (the right-hand side of the system of differential equations) to be cooperative if the off-diagonal terms of the Jacobian matrix $\frac{\partial f}{\partial z}$ are all larger than or equal to zero. Furthermore, we define the vector field $\underline{f}$ to be reducible if the components of the state vector $\underline{z}(\underline{t})$ can be partitioned into two proper subsets for which the systems of differential equations are independent of each other. The vector field $\underline{f}$ is irreducible if it is not reducible.

With these definitions, Hirsch´s sufficient conditions take the following form:

Theorem 2. Assume that the vector field $\underline{f}$ is cooperative and irreducible. Then the resulting flow ϕ is strongly monotone.

We proceed now to apply these results to the system of differential equations (5.2.15), (5.2.18), (5.2.19). We define a set V by

$$V = \{(z_1,z_2,z_3): 0<z_1<1+\mu_2/k_2,\ 0<z_2<1,\ 0<z_3<1+\mu_2/k_2\}. \quad (5.2.32)$$

It is readily shown that no boundary point of the set $\underline{V}$ is an egress point. Thus, if $\underline{v}$ lies on the boundary surface of $\underline{V}$ where $\underline{z}_1 = 0$, then a normal to this boundary pointing into the set $\underline{V}$ is given by $\underline{N} = (1,0,0)$. A tangent to the orbit of the solution of the initial value problem at $\underline{v}$ is given by $\underline{T} = (\mu_1 z_3, -\mu_2 z_2, (\underline{k}_2+\mu_2)(z_2-z_3))$. Hence the inner product of these two vectors is equal to $\underline{N}\cdot\underline{T} = \mu_1 z_3 \geq 0$. A similar treatment of the remaining five boundary surfaces of $\underline{V}$ shows that no boundary point of $\underline{V}$ is an egress point. We conclude therefore that the set $\underline{V}$ is positively invariant.

It follows from the expression (5.2.22) for the Jacobian matrix that the off-diagonal terms are larger than or equal to zero throughout the set $\underline{V}$. Hence the vector field $\underline{f}$ is cooperative. The irreducibility of $\underline{f}$ follows from the fact that no proper subset of the state variables exists that is not influenced by the state variables in the complement of the subset; see equations (5.2.15), (5.2.18), (5.2.19).

By applying Theorem 2 we conclude that the flow $\underline{f}$ is strongly monotone. The set $\underline{V}$ is clearly compact, and it was indicated above how one can prove that it is positively invariant. Since the number of critical points is finite, we can apply Theorem 1 to conclude that the only biologically meaningful solutions of the system of differential equations are those that approach a stationary point as time approaches infinity.

This concludes the discussion of global results. The rest of the analysis of the Macdonald schistosomiasis model is essentially identical to the development in Section 5.1. A detailed discussion and analysis of Macdonald's model is given by Nåsell (1977a).

5.3 A Schistosomiasis Model with Snail Latency and Differential Mortality

In this section we study a transmission model that treats the infection of snails in a more realistic manner than in the previous section. We shall allow for differential mortality and for recovery of infected snails. We present an extension of the results given by Nåsell (1976). The host model used for the snail phase is the last one treated in Section 2.7, while the host model for the human phase is the superinfection process with monogamous mating of Section 2.4. Thirteen basic parameters are needed to establish the model. They are defined as follows:

$\underline{N}_1$ = the number of human hosts;
μ_1 = the death rate per sexually mature parasite;
λ_1 = the egg-laying rate per mated female parasite;
π_1 = the probability for each cercaria to infect a given human being;
$\underline{N}_2$ = the maximum number of snails;
μ_S = the death rate per susceptible snail;
μ_L = the death rate per latent snail;
μ_I = the death rate per infected snail;
μ_R = the death rate per recovered snail;
$\underline{k}_L$ = the rate at which latent snails become infected;
$\underline{k}_I$ = the rate at which infected snails recover;
λ_2 = the rate of cercarial shedding per infected snail;
π_2 = the probability for each egg to infect a given snail.

The model takes the form of a collection of $2\underline{N}_1$ Markov chains and four deterministic functions as follows:

$$\{F^{(1)}(t), M^{(1)}(t), \ldots, F^{(N_1)}(t), M^{(N_1)}(t), S_2(t), L_2(t), I_2(t), R_2(t)\}, \qquad t \geq 0.$$

As in the two previous models, the two Markov chains $\underline{F}^{(k)}$ and $\underline{M}^{(k)}$ associated with each host $\underline{k}$ correspond to the number of female and male parasites. The functions $\underline{S}_2$, $\underline{L}_2$, $\underline{I}_2$, $\underline{R}_2$ give the number of snails that are susceptible, latent, infected, and recovered, respectively. It follows from (2.2.32) that the expected number of parasites per human host $\underline{x}_1(\underline{t})$ satisfies the differential equation

$$x_1' = h_1(t) - \mu_1 x_1. \tag{5.3.1}$$

The functions $\underline{S}_2$, $\underline{L}_2$, $\underline{I}_2$, ,$\underline{R}_2$ satisfy the system of differential equations (2.7.15)-(2.7.18). Rewritten with the notation introduced for the present model, they have the form

$$S_2' = \mu_S N_2 - h_2(t) S_2 - \mu_S S_2, \tag{5.3.2}$$

$$L_2' = h_2(t) S_2 - k_L L_2 - \mu_L L_2, \tag{5.3.3}$$

$$I_2' = k_L L_2 - k_I I_2 - \mu_I I_2, \tag{5.3.4}$$

$$R_2' = k_I I_2 - \mu_R R_2. \tag{5.3.5}$$

The expressions for the infection rates $\underline{h}_1(\underline{t})$ and $\underline{h}_2(\underline{t})$ are found to be

$$h_1(t) = \pi_1 \lambda_2 I_2(t) \tag{5.3.6}$$

and

$$h_2(t) = \tfrac{1}{2}\pi_2 \lambda_1 N_1 \Psi(x_1(t)). \tag{5.3.7}$$

We note that the number of recovered snails $\underline{R}_2$ does not affect the transmission of the infection. It is therefore deleted in the following system of equations. A consequence of this is that the transmission is unaffected by the death rate μ_R per recovered snail. The equation for $\underline{R}_2$ is, however, needed if one wants to evaluate the proportion of snails that are infected since the total number of snails is not constant in this model.

By inserting the infection rates from (5.3.6), (5.3.7) into the system of equations (5.3.1)-(5.3.4) we find that

$$x_1' = \pi_1 \lambda_2 I_2 - \mu_1 x_1, \tag{5.3.8}$$

$$S_2' = \mu_S (N_2 - S_2) - \tfrac{1}{2}\pi_2 \lambda_1 N_1 \Psi(x_1) S_2, \tag{5.3.9}$$

$$L_2' = \frac{1}{2}\pi_2\lambda_1 N_1 \Psi(x_1) S_2 - \nu_L L_2, \tag{5.3.10}$$

$$I_2' = k_L L_2 - \nu_I I_2. \tag{5.3.11}$$

Here we have introduced the two parameters ν_L and ν_I as follows:

$$\nu_L = k_L + \mu_L, \tag{5.3.12}$$

$$\nu_I = k_I + \mu_I. \tag{5.3.13}$$

Thus, the number of parameters has been reduced from twelve to eleven. We proceed to identify suitable state variables that are free of quasi-dimension. At the same time we shall introduce two transmission factors as functions of the eleven parameters that appear in the above equations. A simple way to find the transmission factors is to study the critical points of the system of equations (5.3.8)-(5.3.11). After some elementary evaluations we find that the $\underline{x}_1$-component $\bar{\underline{x}}_1$ of any critical point satisfies the relation

$$\frac{\pi_2\lambda_1 N_1}{\mu_S}\left(\frac{\pi_1\lambda_2 k_L \mu_S N_2}{\mu_1\nu_I\nu_L} - \bar{x}_1\right)\Psi(\bar{x}_1) = 2\bar{x}_1. \tag{5.3.14}$$

This expression is similar in form to (5.1.19) in Section 5.1. The similarity suggests that we introduce two transmission factors $\underline{T}_1$ and $\underline{T}_2$ as follows:

$$T_1 = \frac{\pi_1\lambda_2\bar{N}_2}{\mu_1}, \tag{5.3.15}$$

$$T_2 = \frac{\pi_2\lambda_1 N_1}{\mu_S}, \tag{5.3.16}$$

where

$$\bar{N}_2 = \frac{k_L\mu_S N_2}{\nu_L\nu_I}. \tag{5.3.17}$$

The parameter $\bar{\underline{N}}_2$ is the "effective" number of snails. It is equal to the maximum number of snails $\underline{N}_2$ multiplied by two factors that both are less than one. The first factor is the proportion $\underline{k}_L/\nu_L$ of latent snails that survive the latent period, and the second factor is the ratio μ_S/ν_I of the expected times in the infected state with and without differential mortality and recovery. It is readily verified that $\bar{N}_2$ has the same quasi-dimension as the maximum number of snails N_2, namely S. It follows that T_1 has the quasi-dimension PH^{-1}, which is the same as the quasi-dimension of x_1. In similarity with the treatment in the previous schistosomiasis models we introduce a state variable z_1 free of quasi-dimension by putting

$$z_1 = x_1/T_1. \tag{5.3.18}$$

All three of the state variables S_2, L_2, I_2 have the quasi-dimension S. One way to define corresponding state variables free of quasi-dimension is to divide each of them by the maximum number of snails N_2. However, a little exploration reveals that the differential equations become somewhat simpler if instead we define z_2, z_3, z_4 as follows:

$$z_2 = S_2/N_2, \tag{5.3.19}$$

$$z_3 = \nu_L L_2/(\mu_S N_2). \tag{5.3.20}$$

$$z_4 = I_2/\bar{N}_2. \tag{5.3.21}$$

We also introduce a dimensionless time scale by putting

$$\tau = \mu_S t. \tag{5.3.22}$$

We find from their definitions and equations (5.3.8)-(5.3.11) that the new state variables satisfy the following system of differential equations:

$$z_1' = \alpha_1(z_4 - z_1), \tag{5.3.23}$$

$$z_2' = 1 - z_2 - \tfrac{1}{2}T_2\Psi(T_1 z_1)z_2, \tag{5.3.24}$$

$$z_3' = \alpha_L(\tfrac{1}{2}T_2\Psi(T_1 z_1)z_2 - z_3), \tag{5.3.25}$$

$$z_4' = \alpha_I(z_3 - z_4). \tag{5.3.26}$$

Here, the prime denotes the derivative with respect to τ, and the parameters α_1, α_L, α_I are defined by

$$\alpha_1 = \mu_1/\mu_S, \tag{5.3.27}$$

$$\alpha_L = \nu_L/\mu_S, \tag{5.3.28}$$

$$\alpha_I = \nu_I/\mu_S. \tag{5.3.29}$$

The total number of parameters in the system of equations (5.3.23)-(5.3.26) is five. Any critical point $\bar{\underline{z}}$ of this system of equations can be expressed in the form

$$\bar{z} = (\bar{z}_1,\ 1-\bar{z}_1,\ \bar{z}_1,\ \bar{z}_1), \tag{5.3.30}$$

where the first component $\bar{\underline{z}}_1$ satisfies the equation

$$T_2(1 - \bar{z}_1)\Psi(T_1\bar{z}_1) = 2\bar{z}_1. \tag{5.3.31}$$

The solution of this equation has been studied in detail in Section 5.1. The results of that study allow the following conclusions:

The point $\bar{\underline{z}}^0 = (0,1,0,0)$ is a critical point of the system of equation (5.3.23) - (5.3.26) for all positive values of $\underline{T}_1$ and $\underline{T}_2$. It corresponds to absence of infection from the community. There exist two additional critical points $\bar{\underline{z}}^1$ and $\bar{\underline{z}}^2$ if $\underline{T}_1$ is larger than the threshold value $\underline{\bar{\bar{T}}}_1(\underline{T}_2)$. (The threshold function $\underline{\bar{\bar{T}}}_1$ is the same as the one defined in (5.1.27).) The first component of $\bar{\underline{z}}^i$, $\underline{i} = 1,2$, is equal to $\underline{X}_i(\underline{T}_1,\underline{T}_2)/\underline{T}_1$. The functions $\underline{X}_i$ have been defined and studied in Section 5.1. They can be interpreted as the steady-state values of the expected number of parasites per human host.

These results about threshold functions and critical points are very similar to the corresponding results for the basic schistosomiasis model of Section 5.1. However, the similarity does not extend to cover the local stability properties. In order to study these, we write the system of equations (5.3.23)-(5.3.26) in the form $z' = f(z)$. The Jacobian matrix is then found to be

$$\frac{\partial f}{\partial z}(z) = \begin{pmatrix} -\alpha_1 & 0 & 0 & \alpha_1 \\ -\frac{1}{2}T_1T_2z_2\Psi'(T_1z_1) & -1-\frac{1}{2}T_2\Psi(T_1z_1) & 0 & 0 \\ \frac{1}{2}T_1T_2\alpha_L z_2\Psi'(T_1z_1) & \frac{1}{2}T_2\alpha_L\Psi(T_1z_1) & -\alpha_L & 0 \\ 0 & 0 & \alpha_I & -\alpha_I \end{pmatrix}. \tag{5.3.32}$$

By evaluating this Jacobian matrix at the critical point $\underline{\bar{z}}^0$, we find

$$\frac{\partial f}{\partial z}(\bar{z}^0) = \begin{pmatrix} -\alpha_1 & 0 & 0 & \alpha_1 \\ 0 & -1 & 0 & 0 \\ 0 & 0 & -\alpha_L & 0 \\ 0 & 0 & \alpha_I & -\alpha_I \end{pmatrix}. \tag{5.3.33}$$

This matrix has the four eigenvalues $-\alpha_1$, -1 , $-\alpha_L$, $-\alpha_I$. Since they are all negative we conclude that the critical point $\underline{\bar{z}}^0$ is always asymptotically stable.

The Jacobian matrix evaluated at either of the critical points $\underline{\bar{z}}^i$, $\underline{i} = 1,2$, is simplified by use of relations (5.1.85), (5.1.87). The result is

$$\frac{\partial f}{\partial z}(\bar{z}^i) = \begin{pmatrix} -\alpha_1 & 0 & 0 & \alpha_1 \\ -\dfrac{h(X_i)}{h(X_i)-1} & -\dfrac{T_1}{T_1-X_i} & 0 & 0 \\ \dfrac{\alpha_L h(X_i)}{h(X_i)-1} & \dfrac{\alpha_L X_i}{T_1-X_i} & -\alpha_L & 0 \\ 0 & 0 & \alpha_I & -\alpha_I \end{pmatrix}, \quad i=1,2. \tag{5.3.34}$$

For the further analysis of the eigenvalues of this matrix it is convenient to introduce notation for the arithmetic, geometric, and harmonic means of the three parameters μ_1, α_L, α_I as follows:

$$A = \frac{1}{3}(\alpha_1 + \alpha_L + \alpha_I), \tag{5.3.35}$$

$$G = (\alpha_1 \alpha_L \alpha_I)^{1/3}, \tag{5.3.36}$$

$$H = \frac{3\alpha_1 \alpha_L \alpha_I}{\alpha_1\alpha_L + \alpha_1\alpha_I + \alpha_L\alpha_I} . \tag{5.3.37}$$

It follows that the eigenvalues of the matrix (5.3.34) satisfy the equation

$$\lambda^4 + a\lambda^3 + b\lambda^2 + c\lambda + d = 0, \tag{5.3.38}$$

where

$$a = \frac{T_1}{T_1 - X_i} + 3A, \tag{5.3.39}$$

$$b = 3A\left[\frac{T_1}{T_1 - X_i} + \frac{G^3}{AH}\right], \tag{5.3.40}$$

$$c = \frac{3G^3}{H}\left[\frac{T_1}{T_1 - X_i} - \frac{H/3}{h(X_i)-1}\right], \tag{5.3.41}$$

$$d = G^3\left[\frac{T_1}{T_1 - X_i} - \frac{h(X_i)}{h(X_i)-1}\right]. \tag{5.3.42}$$

The criterion (AIV.21) states that all the roots of equation (5.3.38) have negative real parts if and only if $\underline{a} > 0$, $\underline{b} > 0$, $\underline{d} > 0$, $\underline{f} > 0$, where

$$f = abc - a^2 d - c^2 = ce - a^2 d, \tag{5.3.43}$$

and $\underline{e}$ is defined by

$$e = ab - c. \tag{5.3.44}$$

By using the expressions (5.3.39)-(5.3.41) for $\underline{a}$, $\underline{b}$, $\underline{c}$, we find that $\underline{e}$ can be written in the following form:

$$e = 3A\left[\left(\frac{T_1}{T_1-X_i}\right)^2 + 3A\,\frac{T_1}{T_1-X_i} + \frac{G^3}{3A}\,\frac{1}{h(X_i)-1} + \frac{3G^3}{H}\right]. \tag{5.3.45}$$

For use in the expression for f, we note that d can be written as follows:

$$d = \frac{Hc}{3} - G^3\,\frac{h(X_i)-H/3}{h(X_i)-1}. \tag{5.3.46}$$

By using this expression we find the following form for f:

$$f = cg + G^3a^2\,\frac{h(X_i)-H/3}{h(X_i)-1}, \tag{5.3.47}$$

where

$$g = e - \frac{Ha^2}{3}. \tag{5.3.48}$$

By using expressions (5.3.39) and (5.3.45) for a and e we are led to the following expression for g:

$$g = 3A\left[\left(1-\frac{H}{9A}\right)\left(\frac{T_1}{T_1-X_i}\right)^2 + 3A\left(1-\frac{2H}{9A}\right)\frac{T_1}{T_1-X_i} + \right.$$

$$\left. + \frac{G^3}{3A}\,\frac{1}{h(X_i)-1} + \frac{3G^3}{H}\left(1-\frac{AH^2}{3G^3}\right)\right]. \tag{5.3.49}$$

We proceed to prove that $g > 0$. It is well known that the harmonic, geometric, and arithmetic means of positive quantities satisfy the inequalities

$$H \leq G \leq A; \tag{5.3.50}$$

see e.g. Mitrinovic (1970). Thus the first two terms inside the bracket of (5.3.49) are positive since $X_i < T_1$, $i = 1,2$. The third term inside the bracket is positive since the function h satisfies $h(x) > 2$ for $x > 0$, see (5.1.31). The fourth term inside the bracket is also positive. The inequalities (5.3.50) cannot be used to prove this, since they give no upper bound for the arithmetic mean A. However, Sierpinski (1909) has proved a general inequality relating the harmonic, geometric and arithmetic means of positive quantities that provides such a bound. In the present case with means of three quantities, the inequality takes the form

$$AH^2 \leq G^3. \tag{5.3.51}$$

This inequality proves that the fourth term is positive. Since Sierpinski's inequality appears not to be widely known, a proof is given in Appendix V. This simple proof that $\underline{g} > 0$ shows the advantage of working with the harmonic, geometric, and arithmetic means of α_1, α_L, α_I.

It follows readily from expressions (5.3.39), (5.3.40) and (5.3.45) that $\underline{a} > 0$, $\underline{b} > 0$ and $\underline{e} > 0$ for $\underline{i} = 1,2$. We proceed to consider the sign of $\underline{d}$. By using the definition (5.1.25) of the function $\bar{\underline{T}}_1$, we can express $\underline{d}$ in (5.3.42) as follows:

$$d = G^3 \frac{\bar{T}_1(X_i) - T_1}{(T_1 - X_i)(h(X_i)-1)}. \tag{5.3.52}$$

By using the inequalities (5.1.92), (5.1.93) we find from this expression that $\underline{d} < 0$ for $\underline{i} = 1$ and that $\underline{d} > 0$ for $\underline{i} = 2$. We conclude that the critical point $\bar{\underline{z}}^1$ is unstable and furthermore that the critical point $\bar{\underline{z}}^2$ is asymptotically stable if $\underline{f}$ is positive.

In order to study the sign of $\underline{f}$ with $\underline{i} = 2$ we define two functions $\underline{c}_H$ and $\underline{f}_H$ by the relations

$$c_H(T_1,T_2) = \frac{T_1}{T_1 - X_2} - \frac{H/3}{h(X_2)-1}, \tag{5.3.53}$$

$$f_H(T_1,T_2) = h(X_2') - H/3. \tag{5.3.54}$$

The definitions are meaningful only on and above the threshold since this is the domain where $\underline{X}_2 = \underline{X}_2(\underline{T}_1,\underline{T}_2)$ is defined. The functions $\underline{c}_H$ and $\underline{f}_H$ are related as follows:

$$c_H = \frac{\bar{T}_1(X_2) - T_1}{(T_1-X_2)(h(X_2)-1)} + \frac{f_H}{h(X_2)-1} > \frac{f_H}{h(X_2)-1} . \qquad (5.3.55)$$

We note from (5.3.41) and (5.3.47) that $\underline{c}$ and $\underline{f}$ can be expressed in terms of $\underline{c}_H$ and $\underline{f}_H$ as follows:

$$c = \frac{3G^3 c_H}{H} , \qquad (5.3.56)$$

$$f = cg + \frac{G^3 a^2 f_H}{h(X_2)-1} . \qquad (5.3.57)$$

The functions $\underline{c}_H$ and $\underline{f}_H$ can now be used to establish sufficient conditions for asymptotic stability and instability, respectively, of the critical point $\bar{\underline{z}}^2$. Indeed, we show that $\bar{\underline{z}}^2$ is asymptotically stable if $\underline{f}_H \geq 0$ and that $\bar{\underline{z}}^2$ is unstable if $\underline{c}_H \leq 0$.

To prove this assertion, we note that if $\underline{f}_H \geq 0$, then (5.3.55) shows that $\underline{c}_H > 0$. We can therefore conclude from (5.3.56), (5.3.57) that both $\underline{c}$ and $\underline{f}$ are positive and that therefore $\bar{\underline{z}}^2$ is asymptotically stable.

On the other hand, if $\underline{c}_H \leq 0$, then (5.3.56) shows that $\underline{c} \leq 0$, and it follows from (5.3.55) that $\underline{f}_H < 0$. Thus, (5.3.57) shows that $\underline{f} < 0$. This implies that the critical point $\bar{\underline{z}}^2$ is unstable. In this case, then, both the critical points that exist only above the threshold are unstable. This is a strong indication that the system of equations (5.3.23)-(5.3.26) has periodic solutions. Indeed, a Hopf bifurcation occurs here, since two eigenvalues of the matrix (5.3.34) lie on the imaginary axis when $\underline{i} = 2$ and $\underline{f} = 0$. The existence of periodic solutions in schistosomiasis models has been discussed by Gabriel, Hanisch and Hirsch (1981).

A simple sufficient condition for asymptotic stability of $\bar{\underline{z}}^2$ above the threshold is the following inequality:

$$H \leq 6. \qquad (5.3.58)$$

Assume that this inequality holds. Since $\underline{X}_2 > 0$ and $\underline{h}(\underline{x}) > 2$ for $\underline{x} > 0$ we conclude from (5.3.54) that $\underline{f}_H > 0$, and our above argument shows that this implies that $\bar{\underline{z}}^2$ is asymptotically stable.

Now $\underline{H}$ is defined by (5.3.37) as the harmonic mean of the three quantities $\alpha_1 = \mu_1/\mu_S$, $\alpha_L = (\mu_L + k_L)/\mu_S$ and $\alpha_I = (\mu_I + k_I)/\mu_S$. The paper by Anderson and May (1979) gives a summary of data on the length of the latent period $1/\underline{k}_L$ and on the snail death rates μ_S, μ_L, μ_I. We use their data to make order-of-magnitude estimates of our parameter $\underline{H}$.

The death rate of latent snails is similar to the death rate of susceptible snails, so we take $\mu_L = \mu_S$. The death rate of infected snails is larger than that of susceptible snails. An order-of-magnitude estimate is $\mu_I = 3\mu_S$. The latent period has the same order of magnitude as the mean life span of susceptible snails so we take $\underline{k}_L = \mu_S$. The evidence for recovery of infected snails is inconclusive, but the proportion of infected snails that recover before death tends to be small; we put $\underline{k}_I = 0$. Finally, the mean life span of susceptible snails is of the order of months, while the mean life span of mature schistosomes is of the order of years. Hence we take $\mu_1 \approx 0.1\mu_S$.

By inserting these estimates into (5.3.37) we find that $\underline{H} = 18/65$. This means that the inequality (5.3.58) is satisfied by a good margin;the right-hand side is more than 20 times as large as the left-hand side. Thus, available field data indicate that the condition (5.3.58) for asymptotic stability of the critical point $\bar{\underline{z}}^2$ is satisfied.

It is possible to find two domains above the threshold in the $\underline{T}_1$-$\underline{T}_2$-plane where $\bar{\underline{z}}^2$ is asymptotically stable and unstable, respectively, when the inequality (5.3.58) is not satisfied. We omit details since our finding above indicates that (5.3.58) holds.

We conjecture that the critical point $\bar{\underline{z}}^0$ has a domain of attraction which is small when $\underline{X}_1$ is small, in similarity to the result for the model in Section 5.1. It is left as an open problem to study global stability questions. The results by Hirsch discussed and used in Section 5.2 are not applicable, since the Jacobian matrix (5.3.32) is not cooperative.

The Macdonald schistosomiasis model dealt with in Section 5.2 appears as a special case of the model of this section. With no differential mortality and no recovery we get $\mu_S = \mu_L = \mu_I$ and $\underline{k}_I = 0$. This implies that $\alpha_I = 1$. By using (5.3.37) we find that

$$H = \frac{3\alpha_1\alpha_L}{\alpha_1\alpha_L + \alpha_1 + \alpha_L} . \tag{5.3.59}$$

The harmonic mean $\underline{H}$ is an increasing function of α_1 and α_L. By letting both these parameters tend to infinity we find that

$$H < 3. \tag{5.3.60}$$

Thus (5.3.58) holds for all positive values of α_1 and α_L. This result proves again that $\underline{\bar{z}}^2$ in the Macdonald model is asymptotically stable.

An important conclusion from our study in this section is that the influence of latency, differential mortality, and recovery in the snail population is very simply accounted for by using the expression (5.3.17) for the effective number of snails in the definition (5.3.15) of the transmission factor $\underline{T}_1$. The transmission factor $\underline{T}_2$ is not affected by these phenomena. With transmission factors defined in this way, the study of threshold, worm load and control efficiencies is identical to the study that has been carried out in Section 5.1.

The age-dependence of the snail prevalence is of epidemiological interest. We shall not study it in detail, but we indicate how to proceed with a study of it. We consider a cohort of snails that are exposed to the same conditions as the snails in the model, but where the rate of immigration of new snails is equal to zero. We introduce $\underline{H}_2$ to denote the steady-state infection rate per snail.

More precisely, $\underline{H}_2$ can be described as the steady-state value of the rate at which each susceptible snail becomes latent. By using (5.3.7) and (5.3.16) we find that

$$H_2 = \frac{1}{2}\mu_S T_2 \Psi(X_2). \tag{5.3.61}$$

If the number of susceptible, latent, infected and recovered snails in the cohort are denoted $\underline{S}$, $\underline{L}$, $\underline{I}$, and $\underline{R}$, respectively, we find from equations (3.7.2) - (3.7.5) that

$$S' = -(\mu_S + H_2)S, \tag{5.3.62}$$

$$L' = H_2 S - \nu_L L, \tag{5.3.63}$$

$$I' = k_L L - \nu_I I, \tag{5.3.64}$$

$$R' = k_I I - \mu_R R. \tag{5.3.65}$$

Initially, all the snails have age zero and are susceptible. The initial conditions are therefore $\underline{S}(0) = \underline{S}_0$, $\underline{L}(0) = \underline{I}(0) = \underline{R}(0) = 0$. This initial value problem is linear and can be solved explicitly. The prevalence $\underline{Q}(\underline{s})$ among snails of age $\underline{s}$ is the proportion infected, or

$$Q(s) = \frac{I(s)}{S(s) + L(s) + I(s) + R(s)} . \tag{5.3.66}$$

A study of the age-dependence of this function can be carried out in a straightforward manner.

5.4 A Schistosomiasis Model with Polygamous Mating

The schistosomiasis models dealt with in the previous three sections are all based on the assumption of monogamous mating between male and female schistosomes. We proceed in the present section to consider a model in which polygamous mating is assumed. As host model for the human phase of the life cycle we use the superinfection process with polygamous mating of Section 2.5. The second model of Section 2.6 serves as host model for the snail phase of the life cycle.

By proceeding as in Section 5.1 we can establish a system of differential equations for the two state variables $\underline{x}_1$ and $\underline{p}_2$, where $\underline{x}_1$ is the expected number of parasites per human host and $\underline{p}_2$ is the snail infection probability. By introducing transmission factors through relations (5.1.7) and (5.1.8), we find that the system of differential equations can be written as follows:

$$x_1' = T_1\mu_1 p_2 - \mu_1 x_1, \tag{5.4.1}$$

$$p_2' = \tfrac{1}{2}T_2\mu_2\Psi_P(x_1)(1-p_2) - \mu_2 p_2 , \tag{5.4.2}$$

where

$$\Psi_P(x) = (1 - e^{-x/2})x. \tag{5.4.3}$$

The expected number of mated female parasites per host is given by $\frac{1}{2}\Psi_P(\underline{x}_1)$, see (2.5.4). The subscript $\underline{P}$ is used to indicate that we are assuming polygamous mating in distinction to the monogamous mating of Section 5.1.

The component $\bar{\underline{x}}_1$ of any critical point of system (5.4.1), (5.4.2) is found to satisfy the equation

$$T_2\Psi_P(\bar{x}_1)(T_1 - \bar{x}_1) = 2\bar{x}_1, \tag{5.4.4}$$

which is similar in form to equation (5.1.19) of Section 5.1. This equation serves as starting point for the analysis of the threshold phenomenon. This analysis is very similar to that of Section 5.1, but it is simpler since the function Ψ_P is simpler than the function Ψ. We indicate the main steps in the analysis. We first define an auxiliary function $\underline{h}_P$ by putting

$$h_P(x) = \begin{cases} \dfrac{x\Psi_P'(x)}{x\Psi_P'(x) - \Psi_P(x)}, & x > 0, \\ 2, & x = 0. \end{cases} \tag{5.4.5}$$

Relation (5.1.59) shows that the relation between the functions $\underline{h}_P$ and Ψ_P is the same as that between the functions $\underline{h}$ and Ψ. By using (5.4.3) we find that the function $\underline{h}_P$ can be expressed as as follows:

$$h_P(x) = \begin{cases} 1 + \dfrac{2(e^{x/2} - 1)}{x}, & x > 0, \\ 2, & x = 0. \end{cases} \qquad (5.4.6)$$

In analogy with the definitions (5.1.25) and (5.1.26) we introduce

$$\bar{T}_{1P}(x) = x\, h_P(x), \qquad x \geq 0, \qquad (5.4.7)$$

and

$$\bar{T}_{2P}(x) = \frac{2}{(h_P(x) - 1)\Psi_P(x)}, \qquad x > 0. \qquad (5.4.8)$$

By using (5.4.3) and (5.4.6) we find the following expressions for these two functions:

$$\bar{T}_{1P}(x) = x + 2(e^{x/2} - 1), \quad x \geq 0, \qquad (5.4.9)$$

$$\bar{T}_{2P}(x) = \frac{1}{4\sinh^2(\frac{x}{4})}, \quad x > 0. \qquad (5.4.10)$$

Both of these functions are invertible. We denote their inverses by $\bar{\underline{X}}_{1P}$ and $\bar{\underline{X}}_{2P}$, respectively. An explicit expression in terms of known functions has not been found for $\bar{\underline{X}}_{1P}$, but for $\bar{\underline{X}}_{2P}$ we have

$$\bar{X}_{2P}(T_2) = 4 \operatorname{arcsinh} \frac{1}{2(T_2)^{1/2}}, \quad T_2 > 0. \qquad (5.4.11)$$

Threshold functions are introduced in analogy with (5.1.27) and (5.1.28) as follows:

$$\bar{\bar{T}}_{1P} = \bar{T}_{1P} \circ \bar{X}_{2P}, \qquad (5.4.12)$$

$$\bar{\bar{T}}_{2P} = \bar{T}_{2P} \circ \bar{X}_{1P}. \qquad (5.4.13)$$

Among these two functions we find that the first one has the following explicit form:

$$\bar{\bar{T}}_{1p}(T_2) = 4 \operatorname{arcsinh} \frac{1}{2(T_2)^{1/2}} + \frac{1+(1+4T_2)^{1/2}}{T_2}, \quad T_2 > 0. \qquad (5.4.14)$$

Its asymptotic behaviour for small and large values of $\underline{T}_2$ is found to be as follows:

$$\bar{\bar{T}}_{1p}(T_2) \sim 2/T_2, \quad T_2 \to 0, \qquad (5.4.15)$$

$$\bar{\bar{T}}_{1p}(T_2) \sim 4/(T_2)^{1/2}, \quad T_2 \to \infty. \qquad (5.4.16)$$

A comparison with (5.1.52), (5.1.53) shows that the asymptotic behaviour of the threshold function is the same for polygamous as for monogamous mating.

The model of the present section is highly similar to the one treated in Section 5.1. The origin is the only critical point below the threshold; it is asymptotically stable for all parameter values. Above the threshold there are two additional critical points, of which one is unstable and the other one is asymptotically stable.

The model described in this section can be used to study the influence of the mode of mating of schistosome parasites on the transmission of schistosome infections. Detailed comparisons are made by Nåsell (1978) of three models that are based on assumptions of polygamous mating, monogamous mating, and parthenogenetic egg-laying (where all sexually mature female parasites lay eggs), respectively. The first two of these models are dealt with in the present section and in Section 5.1, respectively. The model with parthenogenetic egg-laying is formulated by a minor variation of the model for hermaphroditic helminthiasis of Section 4.1. The three models are compared with respect to threshold functions, eradication efforts, worm loads, loads of egg-layers, prevalence, incidence, recovery probability, snail prevalence, public health factor, and control efficiencies.

In going from the monogamous via the polygamous to the parthenogenetic case we are dealing with parasites with increasingly more efficient reproductive systems. Associated with this is an increase of the infection level in the community, a lower threshold function, a larger minimum eradication effort, and a lower control efficiency. Thus, parasites with more efficient reproductive systems are harder to eradicate and to control. These findings can all be expected on heuristic grounds.

The analysis by Nåsell (1978) also shows that although there are quantitative differences between monogamous and polygamous mating, the magnitudes by which these two reproductive systems differ are small. The minimum eradication effort with polygamous mating is never larger than 1.32 times the minimum eradication effort with monogamous mating, and the ratios between worm loads, loads of egg-layers, and control efficiencies are all close to one for infections with moderate to high values of the minimum eradication effort. These results lead us to the conclusion that it is not essential from a transmission standpoint to improve the knowledge about the mating behaviour of the parasites.

Comparisons between either mode of mating and the parthenogenetic case give different results. The ratio between minimum eradication efforts is unbounded and ratios between loads of egg-layers and between control efficiencies can become large even when the minimum eradication effort is large. In other words, these indicators can differ by orders of magnitude. This result shows that a schistosomiasis model that ignores mating between male and female parasites may provide a poor approximation.

A model for schistosomiasis with concomitant immunity in human beings has been treated by Nåsell (1977b). The host model used for the human phase of the parasite life cycle is an extension of the host model of Section 2.3. Mating between parasites is assumed to be polygamous as in the present section; this hypothesis leads to more manageable mathematical problems then the alternative hypothesis of monogamous mating. The comparisons made above between the transmission consequences of these two modes of mating indicate that they lead to qualitatively similar results.

Appendix I: The Recovery Probability in the Superinfection Process

In order to investigate the dependence of the recovery probability $\underline{R}(\underline{s},\underline{T})$ on the age $\underline{s}$, we note from (2.2.44) that the function α satisfies the relation

$$\alpha(s+T) = \alpha(s) + \alpha(T) - \alpha(s)\alpha(T). \tag{AI.1}$$

It follows from (2.2.43) that we can write

$$p_0(s+T) = p_0(T)p_0(s)^{1-\alpha(T)} . \tag{AI.2}$$

By insertion of this expression into (2.2.42) we find the following expression for the recovery probability:

$$R(s,T) = p_0(T) \frac{p_0(s)^{1-\alpha(T)} - p_0(s)}{1 - p_0(s)} . \tag{AI.3}$$

For use both here and in Appendix II we define an auxiliary function $\underline{K}_1$ by putting

$$K_1(u,T,A) = \begin{cases} \alpha(T), & u=1, \quad T > 0, \quad A = 1, \\ 1, & u=1, \quad T > 0, \quad A > 1, \\ \dfrac{Au^{1-\alpha(T)} - u}{A - u}, & 0 \leq u < 1, \ T > 0, \ A \geq 1. \end{cases} \tag{AI.4}$$

By applying l'Hospital's rule we can verify that $\underline{K}_1$ is a continuous function of $\underline{u}$ for each fixed value of $\underline{A}$. Differentiation of (AI.4) shows that the partial derivative of $\underline{K}_1$ with respect to $\underline{u}$ equals

$$\frac{\partial}{\partial u} K_1(u,T,A) = \frac{Au^{-\alpha(T)}}{(A-u)^2} K_2(u,T,A), \quad 0<u<1, \ T>0, \ A\geq 1, \tag{AI.5}$$

where the auxiliary function $\underline{K}_2$ is defined by

$$K_2(u,T,A) = A-u^{\alpha(T)} -(A-u)\alpha(T), \quad 0 < u \leq 1, \ T > 0, \ A \geq 1. \tag{AI.6}$$

We note that

$$K_2(1,T,A) = (A-1)(1-\alpha(T)) \geq 0 \qquad \text{(AI.7)}$$

and that

$$\frac{\partial}{\partial u} K_2(u,T,A) = \alpha(T)(1-u^{\alpha(T)-1}) < 0. \qquad \text{(AI.8)}$$

We have thus proved that the function $\underline{K}_2$ is positive on its entire domain of definition. But $\frac{\partial}{\partial u} \underline{K}_1(\underline{u},\underline{T},\underline{A})$ and $\underline{K}_2(\underline{u},\underline{T},\underline{A})$ have the same sign. We conclude therefore that the partial derivative of $\underline{K}_1$ with respect to $\underline{u}$ is positive for $0 < \underline{u} < 1$.

Note now that we can write

$$R(s,T) = p_0(T)K_1(p_0(s),T,1), \quad s \geq 0, \quad T > 0. \qquad \text{(AI.9)}$$

Since $\underline{p}_0(\underline{s})$ is monotonically decreasing in $\underline{s}$, we conclude that the recovery probability $\underline{R}(\underline{s},\underline{T})$ is a monotonically decreasing function of the age $\underline{s}$. Bounds for the recovery probability are found by observing from (AI.9) and (AI.4) that

$$R(0,T) = p_0(T)\,\alpha(T), \qquad \text{(AI.10)}$$

and that

$$R(s,T) \to \frac{p_0(\infty)(1-p_0(T))}{1-p_0(\infty)}, \quad s \to \infty . \qquad \text{(AI.11)}$$

Combining these limiting expressions with the monotonicity of $\underline{R}(\underline{s},\underline{T})$ in $\underline{s}$ leads to the following bounds:

$$p_0(T)\alpha(T) > R(s,T) > \frac{p_0(\infty)(1-p_0(T))}{1-p_0(\infty)}, \quad s > 0, \quad T > 0. \qquad \text{(AI.12)}$$

As a preparation to a study of the dependence of the recovery probability $\underline{R}(\underline{s},\underline{T})$ on $\underline{T}$ we define a function $\underline{f}$ as follows:

$$f(x) = \begin{cases} 1, & x = 0, \\ \dfrac{x}{1-e^{-x}}, & x > 0. \end{cases} \qquad \text{(AI.13)}$$

It is elementary to verify that the function $\underline{f}$ has the following properties:

$$
\begin{aligned}
&1)\ f \text{ is continuous on} && x \geq 0,\\
&2)\ f(x) \sim x, && x \to \infty,\\
&3)\ x < f(x) < x+1, && x > 0,\\
&4)\ \tfrac{1}{2} < f'(x) < 1, && x > 0,\\
&5)\ f(x) \sim 1 + x/2, && x \to 0.
\end{aligned}
\qquad (AI.14)
$$

The proofs of 4) are aided by observing that

$$f'(x) - \frac{1}{2} = \frac{e^{-x}}{(1-e^{-x})^2}(\sinh x - x), \quad x > 0, \qquad (AI.15)$$

and that

$$f'(x) - 1 = \frac{e^{-x}}{(1-e^{-x})^2}(1 - x - e^{-x}), \quad x > 0. \qquad (AI.16)$$

Since $\underline{f}$ is continuous and monotonically increasing, its inverse $\underline{F}$ exists. Clearly we have

$$F : [1, \infty) \to [0, \infty). \qquad (AI.17)$$

We use the functions $\underline{f}$ and $\underline{F}$ to define a function $\underline{T}_0$ as follows:

$$T_0(s;r,H) = -\frac{1}{r}\ln\left(1 - \frac{r}{H}f(rs)\right) \qquad (AI.18)$$

on the domain $\underline{H} > \underline{r}$, $0 \leq \underline{s} < \underline{F}(\underline{H}/\underline{r})/\underline{r}$. We note that a restriction to this domain is necessary in order for the right-hand side of (AI.18) to be real. On the domain where $\underline{T}_0$ is defined we have

$$\alpha(T_0(s)) = \frac{r}{H} f(rs). \qquad (AI.19)$$

We consider now the dependence of the recovery probability $\underline{R}(\underline{s},\underline{T})$ on $\underline{T}$. We note first that we can use (AI.1) to write

$$p_0(s+T) = p_0(s)p_0(T)^{1-\alpha(s)} \tag{AI.20}$$

The expression (2.2.42) for the recovery probability can therefore be rewritten in the form

$$R(s,T) = \frac{p_0(s)}{1-p_0(s)} \left(p_0(T)^{1-\alpha(s)} - p_0(T)\right). \tag{AI.21}$$

We introduce an auxiliary function K_3 by putting

$$K_3(u,s) = u^{1-\alpha(s)} - u, \quad 0 < u < 1, \quad s > 0, \tag{AI.22}$$

and we note that the recovery probability can be expressed with the aid of K_3 through the relation

$$R(s,T) = \frac{p_0(s)}{1-p_0(s)} K_3(p_0(T),s). \tag{AI.23}$$

Differentiation of R with respect to T shows that $\frac{\partial}{\partial T} R(s,T)$ and $\frac{\partial}{\partial u} K_3(p_0(T),s)$ have opposite signs. The partial derivative of K_3 with respect to u is denoted K_4:

$$K_4(u,s) = \frac{\partial}{\partial u} K_3(u,s) = (1 - \alpha(s))u^{-\alpha(s)} - 1. \tag{AI.24}$$

By substituting $p_0(T)$ for the argument u in the function K_4 we get

$$K_4(p_0(T),s) = \exp(\alpha(s)K_5(T,s)) - 1, \tag{AI.25}$$

where the function K_5 is defined by

$$K_5(T,s) = \frac{H}{r}\alpha(T) - \frac{rs}{\alpha(s)} = \frac{H}{r}\alpha(T) - f(rs). \tag{AI.26}$$

We note that $K_4(p_0(T), s)$ and $K_5(T,s)$ have the same sign, and that therefore $\frac{\partial}{\partial T} R(s,T)$ and $K_5(T,s)$ have opposite signs.

If $H \le r$ we use the inequalities $\alpha(T) < 1$, $T \ge 0$, and $f(rs) \ge 1$, $s \ge 0$, to conclude from (AI.26) that

$$K_5(T,s) \le \alpha(T) - f(rs) < 0. \qquad (AI.27)$$

If $H > r$ and $rs \ge F(H/r)$ we note that $f(rs) \ge f(F(H/r)) = H/r$ and hence

$$K_5(T,s) \le \frac{H}{r}\,\alpha(T) - \frac{H}{r} < 0, \quad T \ge 0. \qquad (AI.28)$$

If finally $H > r$ and $0 \le s < F(H/r)/r$, we note by using (AI.19) that

$$K_5(T,s) = \frac{H}{r}\,(\alpha(T) - \alpha(T_0(s))). \qquad (AI.29)$$

Since α is an increasing function we conclude that

$$K_5(T,s) < 0 \quad \text{if} \quad T < T_0(s) \qquad (AI.30)$$

and that

$$K_5(T,s) > 0 \quad \text{if} \quad T > T_0(s). \qquad (AI.31)$$

From (AI.3) we note the following behaviour of the recovery probability $R(s,T)$ for small and large values of T:

$$R(s,T) \to 0, \quad T \to 0, \qquad (AI.32)$$

$$R(s,T) \to e^{-H/r} = p_0(\infty), \quad T \to \infty . \qquad (AI.33)$$

Our results can be summarized as follows:

If $H \le r$ or if $H \ge r$ and $s \ge F(H/r)/r$ then the recovery probability increases monotonically with T toward the value $p_0(\infty)$. If $H > r$ and $s < F(H/r)/r$, then the recovery probability $R(s,T)$ has a maximum for $T = T_0(s;\, r,H)$. This maximum value is larger than $p_0(\infty)$.

Appendix II: The Incidence and the Recovery Probability in the Superinfection Process with Monogamous Mating

By using relations (AI.1) and (2.4.40) we find that the incidence in (2.4.39) can be written in the form

$$I_d(s,T) = (1 - p_{FO}(T))\ (1 - p_{FO}(T)K_1(p_{FO}(s),\ T,\ 2)), \qquad \text{(AII.1)}$$

where the function $\underline{K}_1$ was defined in (AI.4). It was shown in Appendix I that $\underline{K}_1$ is a monotonically increasing function of its first argument. We conclude therefore from (AII.1) that the incidence is an increasing function of the host age $\underline{s}$.

We note that (AI.20) holds with $\underline{p}_0$ replaced by $\underline{p}_{FO}$. By applying the corresponding expression for $\underline{p}_{FO}$ in the expression (2.4.39) for the incidence, we get

$$I_d(s,T) = (1 - p_{FO}(T))\left(1 - \frac{p_{FO}(s)}{2-p_{FO}(s)}\left(2p_{FO}(T)^{1-\alpha(s)} - p_{FO}(T)\right)\right). \qquad \text{(AII.2)}$$

We define an auxiliary function $\underline{K}_6$ by putting

$$K_6(x,a,b) = (1-x)\left(1 - \frac{x(2x^{-a}-1)}{2e^{ab}-1}\right),\quad 0 < a < 1,$$
$$b > 0,\quad e^{-b} \leq x < 1. \qquad \text{(AII.3)}$$

By reference to (2.4.40) we find that the incidence can be expressed in terms of $\underline{K}_6$ as follows:

$$I_d(s,T) = K_6\left(p_{FO}(T),\ \alpha(s),\ \tfrac{H}{2r}\right). \qquad \text{(AII.4)}$$

We proceed to prove that the partial derivative of $\underline{K}_6$ with respect to $\underline{x}$ is negative. Since $\underline{p}_{FO}(\underline{T})$ is a decreasing function of $\underline{T}$, this result will allow us to conclude that the incidence is an increasing function of the length $\underline{T}$ of the time interval over which it is defined.

By differentiating $\underline{K}_6$ with respect to $\underline{x}$ we find that

$$\frac{\partial K_6}{\partial x} = \frac{2x^{-a}(x-(1-x)(1-a)) + 1 - 2x}{2e^{ab} - 1} - 1. \qquad \text{(AII.5)}$$

We define a new function K_7 by subtracting the numerator in the first term of the right-hand side from the denominator and dividing the difference by two. Clearly, $\frac{\partial K_6}{\partial x}$ and K_7 have opposite signs. We shall prove that K_7 is positive. We get the following expression for K_7:

$$K_7(x,a,b) = (e^{ab}+x-1-x^{1-a}) + (1-x)(1-a)x^{-a}. \qquad \text{(AII.6)}$$

The second term in the right-hand side of this expression is positive, so it suffices to prove that the first term is positive. Thus we consider

$$K_8(x,a,b) = e^{ab} + x - 1 - x^{1-a}, \quad 0 < a < 1,$$

$$b > 0, \ e^{-b} \leq x < 1. \qquad \text{(AII.7)}$$

We evaluate this function when x is at the left end-point of its constraining interval. The result is

$$K_8(e^{-b},a,b) = (e^{ab}-1)(1-e^{-b}) > 0. \qquad \text{(AII.8)}$$

Furthermore, the partial derivative of K_8 with respect to x is found to be equal to

$$\frac{\partial K_8}{\partial x} = 1 - (1-a)x^{-a}. \qquad \text{(AII.9)}$$

This derivative is equal to zero when $x = x_0$, where x_0 is equal to

$$x_0 = (1-a)^{1/a}. \qquad \text{(AII.10)}$$

We conclude that the derivative is positive for all x for which the function K_8 is defined if $x_0 \leq e^{-b}$, i.e. if $e^{ab} \leq 1/(1-a)$. By using (AII.8) we conclude in this case that K_8 is positive.

On the other hand, if $x_0 > e^{-b}$, then we conclude that K_8 has a minimum at $x = x_0$. The minimum value is found to be equal to

$$K_8(x_0,a,b) = e^{ab} - 1 - a(1-a)^{(1-a)/a}. \qquad (AII.11)$$

Now $\underline{x}_0 > \underline{e}^{-b}$ implies that $\underline{e}^{ab} > 1/(1-\underline{a})$. We conclude that

$$K_8(x_0,a,b) > \frac{a}{1-a}(1 - (1-a)^{1/a}) > 0. \qquad (AII.12)$$

Thus, in either case, $\underline{K}_8$ is positive. This proves that the incidence is a monotonically increasing function of $\underline{T}$.

Expression (2.4.35) shows that the recovery probability in the dioecious case can be written in the form

$$R_d(s,T) = 1 - (1 - R_F(s,T))^2, \qquad (AII.13)$$

where $\underline{R}_F$ is the monoecious recovery probability for female parasites. The number of females is modelled by an immigration-death process with immigration rate $\underline{H}/2$ and death rate $\underline{r}$. The results concerning the monoecious recovery probability derived in Appendix I lead to the following conclusions:

The recovery probability $\underline{R}_d(\underline{s},\underline{T})$ is a monotonically decreasing function of the host age $\underline{s}$. If $\underline{H} \leq 2\underline{r}$, or if $\underline{H} > 2\underline{r}$ and $\underline{s} \geq \underline{F}(\underline{H}/(2\underline{r}))/\underline{r}$, then the recovery probability $\underline{R}_d(\underline{s},\underline{T})$ increases monotonically with $\underline{T}$ toward the value $2\underline{p}_{FO}(\infty) - \underline{p}_{FO}^2(\infty)$. If $\underline{H} > 2\underline{r}$ and $\underline{s} < \underline{F}(\underline{H}/(2\underline{r}))/\underline{r}$, then the recovery probability $\underline{R}_d(\underline{s},\underline{T})$ has a maximum as function of $\underline{T}$ for $\underline{T} = \underline{T}_0(\underline{s};\underline{r},\underline{H}/2)$.

Appendix III: Control Efficiency Functions for the Ross Malaria Model

A total of fourteen control efficiency functions can be defined for the Ross malaria model. There are two efficiencies of control of the steady-state infection probability $\underline{P}_1$ and three efficiencies of control of each of the steady-state public health factor $\underline{H}_1$, the steady-state prevalence $\underline{Q}(\underline{s})$, the steady-state incidence $\underline{I}(\underline{T})$, and the steady-state recovery probability $\underline{R}(\underline{T})$. Here, $\underline{s}$ denotes the age of the host and $\underline{T}$ denotes the length of the time interval over which incidence and recovery probability are defined. The efficiencies of control of P_1 and H_1 are determined by the two transmission factors T_1 and T_2. The efficiencies of control of $Q(s)$, $I(T)$, and $R(T)$ depend on the three parameters u_1, r_1, and T_2. For brevity, the parameter arguments of the control efficiency functions are deleted.

The efficiencies of control of the steady-state infection probability P_1 were found in (3.1.41) and (3.1.42) to be equal to

$$C_1 = T_1 \frac{\partial P_1}{\partial T_1} = \frac{T_1(T_2+1)}{(T_1+1)^2 T_2} = \frac{T_1}{(T_1+1)K} \tag{AIII.1}$$

and

$$C_2 = T_2 \frac{\partial P_1}{\partial T_2} = \frac{1}{(T_1+1)T_2} , \tag{AIII.2}$$

where K is defined by

$$K = \frac{(T_1+1)T_2}{T_2+1} . \tag{AIII.3}$$

By using (3.1.32) we find that the efficiencies of control of the public health factor H_1 are

$$C_{H_1 u_1} = \frac{u_1}{H_1} \frac{\partial H_1}{\partial u_1} = \frac{T_1 T_2}{T_1 T_2 - 1} , \tag{AIII.4}$$

$$C_{H_1 r_1} = - \frac{r_1}{H_1} \frac{\partial H_1}{\partial r_1} = \frac{1}{T_1 T_2 - 1} , \tag{AIII.5}$$

and

$$C_{H_1 T_2} = \frac{T_2}{H_1} \frac{\partial H_1}{\partial T_2} = \frac{(T_1+1)T_2}{(T_1 T_2 - 1)(T_2+1)} = \frac{K}{T_1 T_2 - 1} . \tag{AIII.6}$$

We use (3.1.32) to express the steady-state prevalence at age s by

$$Q(s) = P_1(1 - e^{-Kr_1 s}). \tag{AIII.7}$$

By defining the function g by

$$g(x,a) = 1 - e^{-x} + axe^{-x}, \quad x \geq 0,\ a \geq -1, \tag{AIII.8}$$

we find that the efficiencies of control of the prevalence $\underline{Q}(\underline{s})$ can be written

$$C_{Qu_1}(s) = u_1 \frac{\partial Q}{\partial u_1} = C_1 g(Kr_1 s, T_1 P_2), \tag{AIII.9}$$

$$C_{Qr_1}(s) = -r_1 \frac{\partial Q}{\partial r_1} = C_1 g(Kr_1 s, -P_2), \tag{AIII.10}$$

and

$$C_{QT_2}(s) = T_2 \frac{\partial Q}{\partial T_2} = C_2 g(Kr_1 s, T_1 P_2). \tag{AIII.11}$$

By using relation (3.1.34) we find that the efficiencies of control of the incidence $\underline{I}(\underline{T})$ take the form

$$C_{Iu_1}(T) = u_1 \frac{\partial I}{\partial u_1} = C_1 g(Kr_1 T, T_1 P_2), \tag{AIII.12}$$

$$C_{Ir_1}(T) = -r_1 \frac{\partial I}{\partial r_1} = C_2 g(Kr_1 T, -P_2), \tag{AIII.13}$$

and

$$C_{IT_2}(T) = T_2 \frac{\partial I}{\partial T_2} = C_2 g(Kr_1 T, T_1 P_2). \tag{AIII.14}$$

We note from (3.1.35) that the recovery probability $\underline{R}(\underline{T})$ can be expressed as follows:

$$R(T) = (1-P_1)\,(1-e^{-Kr_1 T}). \tag{AIII.15}$$

It follows that the efficiencies of control of the recovery probability $\underline{R}(\underline{T})$ are equal to

$$C_{Ru_1}(T) = -u_1 \frac{\partial R}{\partial u_1} = C_1 g(Kr_1 T, -1), \tag{AIII.16}$$

$$C_{Rr_1}(T) = r_1 \frac{\partial R}{\partial r_1} = C_1 g(Kr_1 T, \frac{1}{T_1}), \tag{AIII.17}$$

and

$$C_{RT_2}(T) = -T_2 \frac{\partial R}{\partial T_2} = C_2 g(Kr_1 T, -1). \tag{AIII.18}$$

It is elementary to verify that $\underline{g}$ has the following properties:

1. $\underline{g}(0,\underline{a}) \sim (1+\underline{a})\underline{x}, \ \underline{x} \to 0$,
2. $\underline{g}(\underline{x},\underline{a}) \to 1, \ \ \underline{x} \to \infty$,
3. $\underline{g}$ is a monotonically increasing function of $\underline{x}$ for $-1 \leq \underline{a} \leq 0$,
4. If $\underline{a} > 0$, then $\underline{g}$ as a function of $\underline{x}$ has a unique maximum at $\underline{x} = 1 + 1/\underline{a}$. The maximum value of $\underline{g}$ is $1 + \underline{ae}^{-1-1/a} > 1$.

By using these properties of the function $\underline{g}$, we find the following maximum values for four of the control efficency functions:

$$C_{Qu_1}(s_0) = C_{Iu_1}(s_0) = C_1\left(1 + T_1 P_2 e^{-1/P_1}\right), \qquad \text{(AIII.19)}$$

$$C_{QT_2}(s_0) = C_{IT_2}(s_0) = C_2\left(1 + T_1 P_2 e^{-1/P_1}\right). \qquad \text{(AIII.20)}$$

The maxima of $\underline{C}_{Qu_1}$ and $\underline{C}_{QT_2}$ are taken at $\underline{s} = \underline{s}_0$, and the maxima of $\underline{C}_{Iu_1}$ and $\underline{C}_{IT_2}$ are taken at $\underline{T} = \underline{s}_0$, where

$$s_0 = \frac{T_2+1}{(T_1 T_2 - 1) r_1} = \frac{1}{H_1}. \qquad \text{(AIII.21)}$$

Thus, the larger the steady-state public health factor $\underline{H}_1$ is, the lower is the age $\underline{s}_0$ (the time interval length $\underline{s}_0$) at which the efficiency of control of the prevalence (the incidence) through reductions of $\underline{u}_1$ or $\underline{T}_2$ is maximum.

Furthermore, the function $\underline{C}_{Rr_1}$ takes its maximum at $\underline{T} = \underline{T}_0$, where

$$T_0 = \frac{T_2+1}{T_2 r_1} . \qquad \text{(AIII.22)}$$

Its maximum value is equal to

$$C_{Rr_1}(T_0) = C_1\left(1 + \frac{1}{T_1} e^{-(1+T_1)}\right). \qquad \text{(AIII.23)}$$

Expression (AIII.22) shows that $\underline{T}_0$ is a decreasing function of the transmission factor $\underline{T}_2$; the larger the value of $\underline{T}_2$ is, the smaller is the $\underline{T}$-value at which the control efficiency function $\underline{C}_{Rr_1}$ takes its maximum.

The remaining four control efficiency functions are monotonically increasing functions of $\underline{s}$ or $\underline{T}$. They approach the following values as $\underline{s}$ or $\underline{T}$ approach infinity:

$$C_{Qr_1}(\infty) = C_{Ir_1}(\infty) = C_1, \qquad \text{(AIII.24)}$$

$$C_{Ru_1}(\infty) = C_1, \qquad \text{(AIII.25)}$$

$$C_{RT_2}(\infty) = C_2 \;. \qquad \text{(AIII.26)}$$

The control efficiencies $\underline{C}_1$ and $\underline{C}_2$ are compared in (3.1.45). We proceed to make some additional comparisons of the control efficiencies derived above. By using the inequality $\underline{T}_1\underline{T}_2 > 1$ we find that

$$1 < K < T_1T_2. \qquad \text{(AIII.27)}$$

It follows that

$$C_{H_1u_1} > C_{H_1T_2} > C_{H_1r_1} \qquad \text{(AIII.28)}$$

Thus, a qualitative reversal takes place when the efficiencies of control of the public health factor $\underline{H}_1$ are compared with the efficiencies of control of the infection probability $\underline{P}_1$: Increasing $\underline{r}_1$ is less effective than decreasing $\underline{T}_2$ for controlling $\underline{H}_1$, while the opposite holds for controlling $\underline{P}_1$.

Among the efficiencies of control of the prevalence at any age $\underline{s}$ we find that

$$C_{Qu_1}(s) > C_{Qr_1}(s), \quad C_{Qu_1}(s) > C_{QT_2}(s), \; s > 0. \qquad \text{(AIII.29)}$$

A comparison between the control efficiencies $\underline{C}_{Qr_1}$ and $\underline{C}_{QT_2}$ reveals an age-dependence:

$$C_{Qr_1}(s) > C_{QT_2}(s), \quad s \to \infty, \tag{AIII.30}$$

while

$$C_{Qr_1}(s) < C_{QT_2}(s), \quad s \to 0. \tag{AIII.31}$$

The former of these two inequalities is to be expected, since the prevalence $\underline{Q}(\underline{s})$ at high ages approaches the infection probability $\underline{P}_1$, and the corresponding inequality holds for the efficiencies of control of $\underline{P}_1$.

Results similar to those just quoted for the efficiencies of control of the prevalence hold for the efficiencies of control of the incidence.

Finally, the efficiencies of control of the recovery probability obey the following inequalities:

$$C_{Rr_1}(T) > C_{Ru_1}(T) > C_{RT_2}(T), \; T > 0. \tag{AIII.32}$$

We observe that increasing $\underline{r}_1$ is more effective than decreasing $\underline{u}_1$ or $\underline{T}_2$ for controlling the recovery probability $\underline{R}$.

Comparisons of efficiencies of control of the incidence with efficiencies of control of the recovery probability can be made. They show that

$$C_{Iu_1}(T) > C_{Ru_1}(T), \quad T > 0, \tag{AIII.33}$$

$$C_{Rr_1}(T) > C_{Ir_1}(T), \quad T > 0, \tag{AIII.34}$$

$$C_{IT_2}(T) > C_{RT_2}(T), \quad T > 0. \tag{AIII.35}$$

We observe that an increase of $\underline{r}_1$ has a larger effect on the recovery probability $\underline{R}(\underline{T})$ than on the incidence $\underline{I}(\underline{T})$, while decreases of either $\underline{u}_1$ or $\underline{T}_2$ affect the incidence $\underline{I}(\underline{T})$ more than they affect the recovery probability $\underline{R}(\underline{T})$.

The results discussed here can be used to respond to the question of what indicator is most sensitive to a given control action. The public health factor is not included in this comparison, since it is in itself not directly observable or measurable.

Maximum sensitivity to a lowering of $\underline{u}_1$ is achieved if one observes either the prevalence at age $\underline{s}_0$ or the incidence at any age over a time interval of length $\underline{s}_0$. The control efficiency then takes a value larger than $\underline{C}_1$. This means that the resulting change in prevalence at age $\underline{s}_0$ is larger than the corresponding change in prevalence for old human hosts. Maximum sensitivity to a control action that increases $\underline{r}_1$ is achieved by observing the recovery probability at any age over a time interval of length $\underline{T}_0$. The control efficiency then again takes a value larger than $\underline{C}_1$. Maximum sensitivity to a lowering of $\underline{T}_2$ is achieved by observing either the prevalence at age $\underline{s}_0$ or the incidence at any age over a time interval of length $\underline{s}_0$. The control efficiency then takes a value larger than $\underline{C}_2$.

Appendix IV: Local Stability Results for Equilibrium Solutions of Systems of Differential Equations and of Differential-Difference Equations

We summarize here results concerning stability properties of equilibrium solutions of systems of differential equations and of systems of differential-difference equations. Derivations of the results for differential equations are given in Brauer and Nohel (1969). The results we quote for systems of differential-difference equations are derived by Bellman and Cooke (1963). For an elementary introduction to the theory of differential-difference equations we refer to Driver (1977).

The results will be presented first for a system of differential-difference equations and then for a system of differential equations. We consider first a system of differential-difference equations written in the form

$$z'(t) = f(z(t), z(t-T), \quad T > 0, \quad t \geq 0. \qquad \text{(AIV.1)}$$

Here, $z(t)$ is an n-dimensional vector-valued function of the real variable t. We denote its i'th component by $z_i(t)$, $i = 1,2,\ldots,n$. Furthermore, f is a vector-valued function of two vectors u and v with components u_i and v_i, respecively, $i = 1,2,\ldots,n$. This means that each component f_i of the function f can be written as a function of the $2n$ components of the vectors u and v. We assume that each component of $f(u,v)$ has a continuous partial derivative with respect to each of the $2n$ components $u_1,\ldots,u_n$, $v_1,\ldots,v_n$ of the two vectors u and v.

If the system of equations (AIV.1) is to have a unique solution, then it is necessary that initial values be specified over a t-interval of length T. Thus, we require

$$z(t) = g(t), \quad -T \leq t \leq 0, \qquad \text{(AIV.2)}$$

where g is a given continuous vector-valued function. A function $z(t)$ is said to be a solution of the initial-value problem (AIV.1), (AIV.2) if it satisfies the system of differential-difference equations in (AIV.1) for $t \geq 0$, the initial condition in (AIV.2) on the initial interval $[-T,0]$, and is continuous for $t \geq -T$.

We assume that $\bar{z}$ is an equilibrium solution of (AV.1), i.e. that

$$f(\bar{z},\bar{z}) = 0. \qquad \text{(AIV.3)}$$

The local stability properties of $\bar{z}$ are determined by the roots of a so-called characteristic equation $h(s) = 0$. This equation is determined as follows.

Let $\frac{\partial f}{\partial u}(u,v)$ denote the Jacobian matrix of the transformation that for each fixed vector v maps the vector u into the vector $f(u,v)$. This Jacobian matrix has n rows and n columns and its (ij)'th element equals $\frac{\partial f_i}{\partial u_j}(u,v)$. Similarly, $\frac{\partial f}{\partial v}(u,v)$ denotes the Jacobian matrix of the transformation that for each fixed vector $\underline{u}$ maps the vector $\underline{v}$ into the vector $\underline{f}(\underline{u},\underline{v})$.

The two Jacobian matrices are evaluated at the point where both the arguments $\underline{u}$ and $\underline{v}$ are equal to the equilibrium vactor $\bar{\underline{z}}$. We introduce $\underline{B}_0$ and $\underline{B}_1$ to denote these matrices multiplied by minus one:

$$B_0 = -\frac{\partial f}{\partial u}(\bar{z},\bar{z}), \tag{AIV.4}$$

$$B_1 = -\frac{\partial f}{\partial v}(\bar{z},\bar{z}). \tag{AIV.5}$$

We can now define a characteristic matrix function $\underline{H}(\underline{s})$ by putting

$$H(s) = Is + B_0 + B_1 e^{-Ts}, \tag{AIV.6}$$

where $\underline{I}$ is the unit matrix, and $\underline{s}$ is a complex variable. The characteristic equation is defined by

$$h(s) = \det H(s) = 0, \tag{AIV.7}$$

where the scalar function $\underline{h}(\underline{s})$ is the determinant of the characteristic matrix function $\underline{H}(\underline{s})$. Any solution of this equation is called a characteristic root. The main result can now be stated as follows: The equilibrium solution $\bar{\underline{z}}$ of (AIV.1) is asymptotically stable if all characteristic roots have negative real parts, and it is unstable if at least one characteristic root has positive real part.

The characteristic roots are zeroes of a polynomial in $\underline{s}$ and $\underline{e}^{Ts}$. We shall have occasion to apply the following result for determining when all the roots of a specific such polynomial have negative real parts:

Let

$$k(s) = (s^2 + ps + q)e^{Ts} - r, \tag{AIV.8}$$

where

$$p > 0, \quad q \geq 0, \quad p^2 \geq 2q, \quad T > 0, \quad r > 0. \tag{AIV.9}$$

Then all the roots of the equation $\underline{k}(\underline{s}) = 0$ have negative real parts if and only if $\underline{r} < \underline{q}$.

We turn now to a system of differential equations

$$z'(t) = f(z(t)). \tag{AIV.10}$$

This is a special case of (AIV.1), since $\underline{f}$ depends on only one vector $\underline{z}$. As above we assume that each component of $\underline{f}$ has a continuous partial derivative with respect to each of the components of $\underline{z}$. The initial condition in this case consists in a specification of the value of $\underline{z}(0)$:

$$z(0) = g , \tag{AIV.11}$$

where $\underline{g}$ is a given vector.

We assume that $\underline{\bar{z}}$ is an equilibrium solution of (AIV.10), i.e. that

$$f(\bar{z}) = 0. \tag{AIV.12}$$

The Jacobian matrix evaluated at the equilibrium solution z is denoted by A:

$$A = \frac{\partial f}{\partial z} (\underline{\bar{z}}). \tag{AIV.13}$$

In this case the characteristic matrix function takes the form

$$H(s) = Is - A, \tag{AIV.14}$$

and the characteristic roots satisfy the equation

$$h(s) = \det H(s) = 0, \tag{AIV.15}$$

i.e., they are the eigenvalues of the matrix $\underline{A}$.

The main result can now be stated as follows: The equilibrium solution $\underline{\bar{z}}$ of (AIV.17) is asymptotically stable if all eigenvalues of the matrix $\underline{A}$ have negative real parts, and it is unstable if at least one eigenvalue of $\underline{A}$ has positive real part.

The characteristic equation of the matrix $\underline{A}$ is a polynomial equation with real coefficients. It is desirable to be able to determine from the coefficients of the equation when all roots have negative real parts. A general criterion, due to Routh and Hurwitz, is available. For equations of second, third and fourth degree the results are as follows:

The two roots of the equation

$$x^2 + Ax + B = 0 \tag{AIV.16}$$

have negative real parts if and only if

$$A > 0, \quad B > 0. \tag{AIV.17}$$

The three roots of the equation

$$x^3 + Ax^2 + Bx + C = 0 \tag{AIV.18}$$

have negative real parts if and only if

$$A > 0, \ (B > 0), \ C > 0, \ AB > C. \tag{AIV.19}$$

The four roots of the equation

$$x^4 + Ax^3 + Bx^2 + Cx + D = 0 \tag{AIV.20}$$

have negative real parts if and only if

$$A > 0, \ B > 0, \ (C > 0), \ D > 0, \ (AB > C), \ ABC > A^2D + C^2. \tag{AIV.21}$$

The inequalities in parentheses follow from the others in the same row.

Appendix V: A Proof of Sierpinski s Inequality

For any sequence $a_1, a_2, \ldots, a_k$ of positive numbers we define the arithmetic mean A_k, the geometric mean G_k, and the harmonic mean H_k by

$$A_k = \frac{1}{k} \sum_{i=1}^{k} a_i, \tag{AV.1}$$

$$G_k = \left(\prod_{i=1}^{k} a_i \right)^{1/k}, \tag{AV.2}$$

$$\frac{1}{H_k} = \frac{1}{k} \sum_{i=1}^{k} \frac{1}{a_i}. \tag{AV.3}$$

It is well known that the inequalities

$$H_k \leq G_k \leq A_k, \quad k = 1,2,3,\ldots \tag{AV.4}$$

hold, see e.g. Mitrinovic (1970). We define

$$F_k = H_k^{k-1} A_k / G_k^k \tag{AV.5}$$

and proceed to show that

$$F_{k+1} \leq F_k \leq 1, \quad k = 1,2,3,\ldots . \tag{AV.6}$$

It is readily verified that $F_1 = F_2 = 1$.

The inequalities $F_k \leq 1$ have been proved by Sierpinski (1909). The slightly stronger results in (AV.6) have been proved by Mitrinovic and Vasic (1976).

The idea in the proof is to express the means A_{k+1}, G_{k+1}, H_{k+1} in terms of the means A_k, G_k, H_k and the element a_{k+1}. We put $x = a_{k+1}$ and note from the definitions (AV.1)-(AV.3) that

$$A_{k+1} = \frac{1}{k+1} (kA_k + x), \tag{AV.7}$$

$$G_{k+1}^{k+1} = G_k^k \, x , \tag{AV.8}$$

and

$$H_{k+1} = \frac{(k+1)H_k x}{H_k + kx}. \tag{AV.9}$$

We proceed to study F_{k+1} as a function of x. From (AV.5) and (AV.7) - (AV.9) we find that

$$F_{k+1}(x) = \frac{(k+1)^{k-1} H_k^k}{G_k^k} \frac{(x+kA_k)x^{k-1}}{(kx+H_k)^k} . \tag{AV.10}$$

The derivative of F_{k+1} with respect to x is found to be

$$F'_{k+1}(x) = - \frac{k(k+1)^{k-1} H_k^k}{G_k^k} \frac{(kA_k-H_k)x^{k-2}}{(kx+H_k)^{k+1}} (x-x_0), \quad k \geq 2,$$

where

$$x_0 = \frac{(k-1)A_k H_k}{kA_k - H_k} . \tag{AV.12}$$

The inequality between harmonic and arithmetic means in (AV.4) shows that $x_0 > 0$ and that F_{k+1} has a maximum at $x = x_0$. The maximum value is equal to

$$F_{k+1}(x_0) = F_k \left((k^2-1)A_k/(k^2A_k-H_k)\right)^{k-1} \tag{AV.13}$$

The factor multiplying F_k in the right-hand side of this expression is an increasing function of H_k. The inequality between harmonic and arithmetic means in (AV.4) implies that this factor is always less than or equal to one. Thus (AV.6) has been proved.

REFERENCES

Abramowitz, A. & Stegun, I. A. (1968). Handbook of Mathematical Functions. New York, N.Y.: Dover Publications, Inc.

Anderson, R. M. & May, R. M. (1979). Prevalence of schistosome infections within molluscan populations: observed patterns and theoretical predictions. Parasitology, 79, 63-94.

Bailey, N. T. J. (1964). The Elements of Stochastic Processes with Applications to the Natural Sciences. New York, London, Sidney: John Wiley & Sons, Inc.

Bailey, N. T. J. (1975). The Mathematical Theory of Infectious Diseases and Its Applications. London and High Wycombe: Charles Griffin and Company Ltd.

Bailey, N. T. J. (1982). The Biomathematics of Malaria. London and High Wycombe: Charles Griffin and Company Ltd.

Bellman, R. & Cooke, K.L. (1963). Differential-Difference Equations. New York and London: Academic Press Inc.

Brauer, F. & Nohel, J.A. (1969). The Qualitative Theory of Ordinary Differential Equations. New York and Amsterdam: W.A. Benjamin, Inc.

Braun, M. (1975). Differential Equations and Their Applications. New York, Berlin, Heidelberg: Springer-Verlag.

Chandler, A. C. & Read, C. P. (1961). Introduction to Parasitology. New York, London, Sidney: John Wiley & Sons, Inc.

Cohen, J. E. (1977). Mathematical models of schistosomiasis. Ann. Rev. Ecol. System., 8, 209-233.

Dietz, K., Molineaux, L. & Thomas, A. (1974). A malaria model tested in the African Savannah. Bull. W. H. O., 50, 347-357.

Driver, R.D. (1977). Ordinary and Delay Differential Equations. New York, Heidelberg, Berlin: Springer-Verlag.

Fine, P. E. M. (1975). Superinfection - a problem in formulating a problem. Trop. Dis. Bull., 72, 475-488.

Gabriel, J. P., Hanisch, H. & Hirsch, W. M. (1981). Dynamic equilibria in helminth infections? In: Quantitative Population Dynamics, Chapman, D.G., & Gallucci, V.F. (Editors), Fairland, Maryland: International Co-operative Publishing House, 83-104.

Hirsch, M.W. (1984). The dynamical systems approach to differential equations. Bull. Am. Math. Soc., 11, 1-64.

Jordan, P. & Webbe, G. (1982). Schistosomiasis: Epidemiology, Treatment and Control. London: William Heinemann Medical Books Ltd.

Lin, C.C. & Segel, L.A. (1974). Mathematics Applied to Deterministic Problems in the Natural Sciences. New York: Macmillan Publishing Co, Inc.

Macdonald, G. (1950). The analysis of infection rates in diseases in which superinfection occurs. Trop.Dis.Bull., 47, 907-915.

Macdonald, G. (1952). The analysis of equilibrium in malaria. Trop. Dis.Bull., 49, 813-829.

Macdonald, G. (1957). The Epidemiology and Control of Malaria. London: Oxford Univ. Press.

Macdonald, G. (1965). The dynamics of helminthic infections, with special reference to schistosomes. Trans.Roy.Soc.Trop.Med.Hyg., 59, 489-504.

Macdonald, G. (1973). Dynamics of Tropical Diseases. London, New York, Toronto: Oxford Univ. Press.

Mitrinovic, D.S. (1970). Analytic Inequalities. Berlin, Heidelberg, New York: Springer-Verlag.

Mitrinovic, D. S. & Vasic, P. M. (1976). On a theorem of W. Sierpinsky concerning means. Univ.Beograd. Publ. Electrotechn.Fak., Ser.Mat.Fiz., 544-576, 113-114.

Nåsell, I. (1976). A hybrid model of schistosomiasis with snail latency. Theor.Pop.Biol., 10, 47-69.

Nåsell, I. (1977a). On transmission and control of schistosomiasis, with comments on Macdonald s model. Theor.Pop.Biol., 12, 335-365.

Nåsell, I. (1977b). Schistosomiasis with concomitant immunity. Bull.Int.Stat.Inst., 47(2), 3-21.

Nåsell, I. (1978). Mating models for schistosomes. J.Math.Biol., 6, 21-35.

Nåsell, I. (1984). The role of the breakpoint in schistosomiasis eradiation. In: Proceedings of the Seventh Conference on Probability Theory, Bucharest: Editura Academiei Republicii Socialiste Romania.

Nåsell, I. & Hirsch, W. M. (1972). A mathematical model of some helminthic infections. Comm.Pure Appl.Math., 25, 459-477.

Nåsell, I. & Hirsch, W. M. (1973). The transmission dynamics of schistosomiasis. Comm.Pure Appl.Math., 26, 395-453.

Ross, R. (1909). Report on the Prevention of Malaria in Mauritius. London: Churchill.

Ross, R. (1911). The Prevention of Malaria. London: Murray.

Sierpinski, W. (1909). Sur une inégalité pour la moyenne arithmétique, geométrique et harmonique. Warsch. Sitzungsber., 2, 354-357.

Soni, R.P. (1965). On an inequality for modified Bessel functions. J.Math.Phys., 44, 406-407.

SUBJECT INDEX